2010

TLVs® and BEIs®

Based on the Documentation of the

Threshold Limit Values

for Chemical Substances
and Physical Agents

&

Biological Exposure Indices

ACGIH®

**Defining the Science of
Occupational and Environmental Health®**

Signature Publications

ISBN: 978-1-607260-19-6

Printed in the United States.

ACGIH® is a member-based organization that advances occupational and environmental health. The organization has contributed substantially to the development and improvement of worker health protection. The organization is a professional society, not a government agency.

The *Documentation of the Threshold Limit Values and Biological Exposure Indices* is the source publication for the TLVs® and BEIs® issued by ACGIH®. That publication gives the pertinent scientific information and data with reference to literature sources that were used to base each TLV® or BEI®. For better understanding of the TLVs® and BEIs®, it is essential that the *Documentation* be consulted when the TLVs® or BEIs® are being used. For further information, contact The Science Group, ACGIH®. The most up-to-date list of substances and agents under study by the Committees is available at www.acgih.org/TLV/Studies.htm.

Comments, suggestions, and requests for interpretations or technical information should be directed to The Science Group at the address below or to the following E-mail address: science@acgih.org. To place an order, visit our website at www.acgih.org/store, contact Customer Service at the address or phone number below, or use the following E-mail address: customerservice@acgih.org.

Help ensure the continued development of TLVs® and BEIs®. Make a tax deductible donation to the FOHS Sustainable TLV®/BEI® Fund today!

http://www.fohs.org/SusTLV-BEIPrgm.htm

ACGIH®
1330 Kemper Meadow Drive
Cincinnati, OH 45240-4148
Telephone: 513-742-2020; Fax: 513-742-3355
www.acgih.org

In the event significant errata are required, they
will be listed on the ACGIH® website at
http://www.acgih.org/TLV/.

TABLE OF CONTENTS

Biological Exposure Indices

Physical Agents

Biologically Derived Airborne Contaminants

STATEMENT OF POSITION REGARDING
THE TLVs® AND BEIs®

The American Conference of Governmental Industrial Hygienists (ACGIH®) is a private, not-for-profit, nongovernmental corporation whose members are industrial hygienists or other occupational health and safety professionals dedicated to promoting health and safety within the workplace. ACGIH® is a scientific association. ACGIH® is not a standards-setting body. As a scientific organization, it has established committees that review the existing published, peer-reviewed scientific literature. ACGIH® publishes guidelines known as Threshold Limit Values (TLVs®) and Biological Exposure Indices (BEIs®) for use by industrial hygienists in making decisions regarding safe levels of exposure to various chemical and physical agents found in the workplace. In using these guidelines, industrial hygienists are cautioned that the TLVs® and BEIs® are only one of multiple factors to be considered in evaluating specific workplace situations and conditions.

Each year, ACGIH® publishes its TLVs® and BEIs® in a book. In the introduction to the book, ACGIH® states that the TLVs® and BEIs® are guidelines to be used by professionals trained in the practice of industrial hygiene. The TLVs® and BEIs® are not designed to be used as standards. Nevertheless, ACGIH® is aware that in certain instances the TLVs® and the BEIs® are used as standards by national, state, or local governments.

Governmental bodies establish public health standards based on statutory and legal frameworks that include definitions and criteria concerning the approach to be used in assessing and managing risk. In most instances, governmental bodies that set workplace health and safety standards are required to evaluate health effects, economic and technical feasibility, and the availability of acceptable methods to determine compliance.

ACGIH® TLVs® and BEIs® are not consensus standards. Voluntary consensus standards are developed or adopted by voluntary consensus standards bodies. The consensus standards process involves canvassing the opinions, views, and positions of all interested parties and then developing a consensus position that is acceptable to these parties. While the process used to develop a TLV® or BEI® includes public notice and requests for all available and relevant scientific data, the TLV® or BEI® does not represent a consensus position that addresses all issues raised by all interested parties (e.g., issues of technical or economic feasibility). The TLVs® and BEIs® represent a scientific opinion based on a review of existing peer-reviewed scientific literature by committees of experts in public health and related sciences.

ACGIH® TLVs® and BEIs® are health-based values. ACGIH® TLVs® and BEIs® are established by committees that review existing published and peer-reviewed literature in various scientific disciplines (e.g., industrial hygiene, toxicology, occupational medicine, and epidemiology). Based on the available information, ACGIH® formulates a conclusion on the level of exposure that the typical worker can experience without adverse health effects. The TLVs® and BEIs® represent conditions under which ACGIH® believes that nearly all workers may be repeatedly exposed without adverse health effects. They are not fine lines between safe and dangerous exposures, nor are they a relative index of toxicology. The TLVs® and BEIs® are not quantitative estimates of risk at different exposure levels or by different routes of exposure.

Since ACGIH® TLVs® and BEIs® are based solely on health factors, there is no consideration given to economic or technical feasibility. Regulatory agencies should not assume that it is economically or technically feasible for an industry or employer to meet TLVs® or BEIs®. Similarly, although there are usually valid methods to measure workplace exposures at the TLVs® and BEIs®, there can be instances where such reliable test methods have not yet been validated. Obviously, such a situation can create major enforcement difficulties if a TLV® or BEI® was adopted as a standard.

ACGIH® does not believe that TLVs® and BEIs® should be adopted as standards without full compliance with applicable regulatory procedures, including an analysis of other factors necessary to make appropriate risk management decisions. However, ACGIH® does believe that regulatory bodies should consider TLVs® or BEIs® as valuable input into the risk characterization process (hazard identification, dose-response relationships, and exposure assessment). Regulatory bodies should view TLVs® and BEIs® as an expression of scientific opinion.

ACGIH® is proud of the scientists and the many members who volunteer their time to work on the TLV® and BEI® Committees. These experts develop written *Documentation* that includes an expression of scientific opinion and a description of the basis, rationale, and limitations of the conclusions reached by ACGIH®. The *Documentation* provides a comprehensive list and analysis of all the major published peer-reviewed studies that ACGIH® relied upon in formulating its scientific opinion. Regulatory agencies dealing with hazards addressed by a TLV® or BEI® should obtain a copy of the full written *Documentation* for the TLV® or BEI®. Any use of a TLV® or BEI® in a regulatory context should include a careful evaluation of the information in the written *Documentation* and consideration of all other factors as required by the statutes which govern the regulatory process of the governmental body involved.

- *ACGIH® is a not-for-profit scientific association.*

- *ACGIH® proposes guidelines known as TLVs® and BEIs® for use by industrial hygienists in making decisions regarding safe levels of exposure to various hazards found in the workplace.*

- *ACGIH® is not a standard-setting body.*

- *Regulatory bodies should view TLVs® and BEIs® as an expression of scientific opinion.*

- *TLVs® and BEIs® are not consensus standards.*

- *ACGIH® TLVs® and BEIs® are based solely on health factors; there is no consideration given to economic or technical feasibility. Regulatory agencies should not assume that it is economically or technically feasible to meet established TLVs® or BEIs®.*

- *ACGIH® believes that TLVs® and BEIs® should NOT be adopted as standards without an analysis of other factors necessary to make appropriate risk management decisions.*

- *TLVs® and BEIs® can provide valuable input into the risk characterization process. Regulatory agencies dealing with hazards addressed by a TLV® or BEI® should review the full written Documentation for the numerical TLV® or BEI®.*

ACGIH® is publishing this Statement in order to assist ACGIH® members, government regulators, and industry groups in understanding the basis and limitations of the TLVs® and BEIs® when used in a regulatory context. This Statement was adopted by the ACGIH® Board of Directors on March 1, 2002.

TLV®/BEI® DEVELOPMENT PROCESS: AN OVERVIEW

Provided below is an overview of the ACGIH® TLV® and BEI® development process. Additional information is available on the ACGIH® website (www.acgih.org). Please also refer to the attached Process Flowchart (Figure 1).

1. **Under Study:** Each committee determines its own selection of chemical substances or physical agents for its Under Study list. A variety of factors is used in this selection process, including prevalence, use, number of workers exposed, availability of scientific data, existence/absence of a TLV® or BEI®, age of TLV® or BEI®, input from the public, etc. The public may offer input to any TLV® or BEI® committee by e-mail to science@acgih.org.

   When a substance or agent is selected for the development of a TLV® or BEI® or for review of an adopted value, the appropriate Committee places it on its Under Study list. This list is published each year by February 1 on the ACGIH® website (www.acgih.org/TLV/Studies.htm), in the ACGIH® Annual Reports, and later in the annual *TLVs® and BEIs®* book. In addition, the Under Study list is updated by July 31 into a two-tier list.

   - Tier 1 entries indicate which chemical substances and physical agents **may** move forward as an NIC or NIE in the upcoming year, based on their status in the development process.
   - Tier 2 consists of those chemical substances and physical agents that **will not** move forward, but will either remain on or be removed from the Under Study list for the next year.

   This updated list will remain in two-tiers for the balance of the year. ACGIH® will continue this practice of updating the Under Study list by February 1 and establishing the two-tier list by July 31 each year.

   The Under Study lists published in the ACGIH® Annual Reports and the annual *TLVs® and BEIs®* book are current as of January 1. All updates to the Under Study lists and publication of the two-tier lists are posted on the ACGIH® website (http://www.acgih.org/TLV/Studies.htm).

   The Under Study list serves as a notification and invitation to interested parties to submit substantive data and comments to assist the Committee in its deliberations. Each Committee considers only those comments and data that address the health science, not economic or technical feasibility. Comments must be accompanied by copies of substantiating data, preferably in the form of peer-reviewed literature. Should the data be from unpublished studies, ACGIH® requires written authorization from the owner of the studies granting ACGIH® permission to (1) <u>use</u>, (2) <u>cite</u> within the *Documentation*, and (3) upon request from a third party, <u>release</u> the information. All three permissions must be stated/covered in the written authorization. (See endnote for a sample permission statement.) Electronic submission of all information to the ACGIH® Science Group at science@acgih.org greatly increases the ease and efficiency with which the Committee can consider the comments or data.

2. **Draft *Documentation*:** One or more members of the appropriate Committee are assigned the task of collecting information and data from the scientific literature, reviewing results of unpublished studies submitted for review, and developing a draft TLV® or BEI® *Documentation*. The draft *Documentation* is a critical evaluation of the scientific literature relevant to recommending a TLV® or BEI®; however, it is not an exhaustive or broad-based critical review of the scientific literature. Particular emphasis is given to papers that address minimal or no adverse health effect levels in exposed animals or workers, that deal with the reversibility of such effects, or in the case of a BEI®, that assess chemical uptake and provide applicable determinant(s) as an index of uptake. Human data, when available, are given special emphasis. This draft *Documentation*, with its proposed TLV® or BEI®, is then reviewed and critiqued by additional Committee members, and eventually by the full Committee. This often results in several revisions to the draft *Documentation* before the full Committee accepts the proposed TLV® or BEI® and *Documentation*. The draft *Documentation* is not available to the public through this stage of the development process and is not released until it is at the Notice of Intended Changes (NIC) stage. Authorship of the *Documentation* is not disclosed.

3. **Notice of Intended Changes (NIC):**

   *[**Notice of Intent to Establish (NIE):** The physical agents section of the TLVs® and BEIs® book also uses the term Notice of Intent to Establish (NIE) in addition to NIC. An NIE follows the same development process as an NIC. For purposes of this process overview, only the term NIC is used.]*

   When the full Committee accepts the draft *Documentation* and its proposed TLV® or BEI®, the *Documentation* and proposed values are then recommended to the ACGIH® Board of Directors for ratification as an NIC. If ratified, each proposed TLV® or BEI® is published as an NIC in the *Annual Reports of Committees on TLVs® and BEIs®*, which is published in the ACGIH® member newsletter, *Today! Online* and is also available online for purchase at http://www.acgih.org/store. At the same time, the draft *Documentation* is made available through ACGIH® Customer Service or online at http://www.acgih.org/store. All information contained in the Annual Reports is integrated into the annual *TLVs® and BEIs®* book, which is usually available to the general public in February or March of each year. The proposed TLV® or BEI® is considered a trial limit by ACGIH® for approximately one year following the NIC ratification by the ACGIH® Board of Directors. Interested parties, as well as ACGIH® members, are invited to provide data and substantive comments, preferably in the form of peer-reviewed literature, on the proposed TLVs® or BEIs® contained in the NIC. Should the data be from unpublished studies, ACGIH® requires written authorization from the owner of the studies granting ACGIH® permission to (1) use, (2) cite within the *Documentation*, and (3) upon request from a third party, release the information. All three permissions must be stated/covered in the written authorization. (See endnote for a sample permission statement.) The most effective and helpful comments are those that address specific points within the draft *Documentation*. Changes or updates are made to the draft *Documentation* as necessary. If the Committee finds or receives substantive data that change its scientific opinion regarding an NIC TLV® or BEI®, and possibly change its proposed TLV® or BEI® values or notations, the Committee may revise the proposal(s) and recommend to the ACGIH® Board of Directors that it be retained on the NIC.

Important Notice: The comment period for an NIC or NIE draft *Documentation* and its respective TLV(s)®, notation(s), or BEI(s)® is limited to a firm 6-month period, running from February 1 to July 31 of each year. ACGIH® restructured the comment period effective January 1, 2007 to ensure all comments are received by ACGIH® in time for full consideration by the appropriate Committee before its fall meeting. Because of the time required to review, evaluate, and consider comments during the fall meetings, any comments received after the July 31 deadline will not be considered in that year's committee deliberations regarding the outcome for possible adoption of an NIC or NIE. As general practice, ACGIH® reviews all comments regarding chemical substances and physical agents on the Under Study list, as well as NICs or NIEs, or currently adopted TLV(s)® or BEI(s)®. All comments received after July 31 will be fully considered in the following year. Draft *Documentation* will be available for review during the full 6-month period.

When submitting comments, ACGIH® requires that the submission be limited to 10 pages in length, including an executive summary. The submission may include appendices of citable material not included as part of the 10-page limit. It would be very beneficial to structure comments as follows:

A. **Executive Summary** – *Provide an executive summary with a limit of 250 words.*

B. **List of Recommendations/Actions** – *Identify, in a vertical list, specific recommendations/actions that are being requested.*

C. **Rationale** – *Provide specific rationale to justify each recommendation/action requested.*

D. **Citable Material** – *Provide citable material to substantiate the rationale.*

The above italicized procedure is requested to permit ACGIH® to more efficiently and productively review comments.

4. **TLV®/BEI® and Adopted *Documentation*:** If the Committee neither finds nor receives any substantive data that change its scientific opinion regarding an NIC TLV® or BEI®, the Committee may then approve its recommendation to the ACGIH® Board of Directors for adoption. Once approved by the Committee and subsequently ratified by the Board, the TLV® or BEI® is published as adopted in the *Annual Reports of the Committees on TLVs® and BEIs®* and in the annual *TLVs® and BEIs®* book, and the draft TLV® or BEI® *Documentation* is finalized for formal publication.

5. **Withdraw from Consideration:** At any point in the process, the Committee may determine not to proceed with the development of a TLV® or BEI® and withdraw it from further consideration. Substances or physical agents that have been withdrawn from consideration can be reconsidered by placement on the Under Study List (step 1 above).

There are *several important points* to consider throughout the above process:

i. The appropriate method for an interested party to contribute to the TLV® and BEI® process is through the submission of literature that is peer-reviewed and public. ACGIH® strongly encourages interested parties to publish their studies, and not to rely on unpublished studies as their input to the TLV® and BEI® process. Also, the best time to submit comments to ACGIH® is in the early stages of the TLV® and BEI® development process, preferably while the substance or agent is on the Under Study list.

ii. An additional venue for presentation of new data is an ACGIH®-sponsored symposium or workshop that provides a platform for public discussion and scientific interpretation. ACGIH® encourages input from external parties for suggestions on symposium topics, including suggestions about sponsors, speakers and format. ACGIH® employs several criteria to determine the appropriateness of a symposium. A key criterion is that the symposium must be the most efficient format to present the Committee with information that will assist in the scientific judgment used for writing the *Documentation* and in setting the respective TLVs® or BEIs®. A symposium topic should be suggested while the substance/agent is Under Study, as symposia require considerable time, commitment, and resources to develop. Symposium topic suggestions submitted while a substance is on the NIC will be considered, but this is usually too late in the decision-making process. A symposium topic will not be favorably considered if its purpose is to provide a forum for voicing opinions about existing data. Rather, there must be ongoing research, scientific uncertainty about currently available data, or another scientific reason for the symposium. Symposium topic suggestions should be sent to the ACGIH® Science Group (science@acgih.org).

iii. ACGIH® periodically receives requests from external parties to make a presentation to a committee about specific substances or issues. It is *strictly by exception* that such requests are granted. While there are various reasons for this position, the underlying fact is that the Committee focuses on data that have been peer-reviewed and published and not on data presented in a private forum. A committee may grant a request when the data is significantly new, has received peer review, is the best vehicle for receipt of the information, and is essential to the committee's deliberations. The presentation is not a forum to voice opinions about existing data. In order for a committee to evaluate such a request, the external party must submit a request in writing that, at a minimum, addresses the following elements: (a) a detailed description of the presentation; (b) a clear demonstration of why the information is important to the Committee's deliberations; and (c) a clear demonstration of why a meeting is the necessary method of delivery. This request must be sent to the ACGIH® Science Group (science@acgih.org).

Also, the Committee may initiate contact with outside experts (a) to meet with the Committee to discuss specific issues or to obtain additional knowledge on the subject, and (b) to provide written input or review of a *Documentation*. This is only done on an as needed basis, and not as a routine practice.

iv. ACGIH® does *not* commit to deferring consideration of a new or revised TLV® or BEI® pending the outcome of proposed or ongoing research.

Important dates to consider throughout each calendar year of the TLV®/BEI® Development Process:

First Quarter:

- The TLV®/BEI® Annual Reports and the *TLVs® and BEIs®* book are published.

Year Round:

- Public comments are accepted.*

- Committees meet.

* Note: It is recommended that comments be submitted as early as practical, and preferably no later than July 31st to allow sufficient time for their proper consideration/review. This is essential for an NIC or NIE TLV®/BEI®.

Important Notice: The comment period for an NIC or NIE draft *Documentation* and its respective TLV(s)®, notation(s), or BEI(s)®, is limited to a firm 6-month period, running from February 1 to July 31 of each year. ACGIH® restructured the comment period effective January 1, 2007 to ensure all comments are received by ACGIH® in time for full consideration by the appropriate Committee before its fall meeting.

Third Quarter:

- Two-tier Under Study list published on website (http://www.acgih.org/TLV/Studies.htm).

Fourth Quarter: **

- TLV®/BEI® Committees vote on proposed TLVs®/BEIs® for NIC or final adoption.

- ACGIH® Board of Directors ratifies TLV®/BEI® Committee recommendations.

** Note: These actions typically occur early in the fourth quarter, but may occur during other periods of the quarter or year.

Endnote: Sample permission statement granting ACGIH® authorization to use, cite, and release unpublished studies:

[Name], [author or sponsor of the study*] grants permission to ACGIH® to use and cite the documents listed below, and to fully disclose them to parties outside of ACGIH® upon request.

Permission to disclose the documents includes permission to make copies as needed.

Example: Joseph D. Doe, PhD, co-author of the study, grants permission to ACGIH® to use and cite the document listed below, and to fully disclose this document to parties outside of ACGIH®. Permission to disclose the document includes permission to make copies as needed.

"Effects of Quartz Status on Pharmacokinetics of Intratracheally Instilled Cristobalite in Rats, March 21, 2003."

*This statement must be signed by an individual authorized to give this permission, and should include contact information such as title and address.

Last Revised January 31, 2008

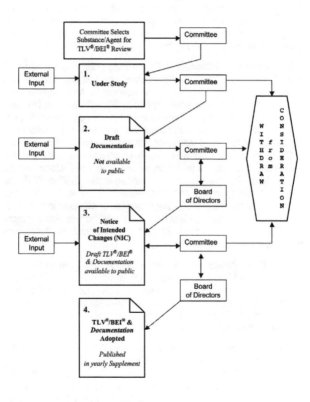

FIGURE 1. The TLV®/BEI® Development Process Flow Chart.

December 20, 2004

ONLINE TLV® AND BEI® RESOURCES

In an effort to make the threshold limit values (TLVs®) and biological exposures indices (BEIs®) guideline establishment process more transparent, and to assist ACGIH® members, government regulators, and industry groups in understanding the basis and limitations of the TLVs® and BEIs®, ACGIH® has an online TLV®/BEI® Resources Section on its website at www.acgih.org/TLV/.

The TLV®/BEI® Resources Section is divided into eight categories, each containing clear and concise information. The categories are:

- **Conflict of Interest Policy** — applies to the Board of Directors, Committee Chairs, and Committee members (including consultant members), and safeguards the integrity and credibility of ACGIH® programs and activities. The Policy, as well as ACGIH®'s oversight and review, each play an important part in the protection of ACGIH®'s programs and activities from inappropriate influences (www.acgih.org/TLV/COIPolicy.htm).

- **Notice of Intended Changes (NIC)** — a listing of the proposed actions of the TLV®-CS, TLV®-PA, and BEI® Committees. This Notice provides an opportunity for public comment. Values remain on the NIC for approximately one year after they have been ratified by ACGIH®'s Board of Directors. The proposals should be considered trial values during the period they are on the NIC. If the Committee neither finds nor receives any substantive data that change its scientific opinion regarding an NIC TLV® or BEI®, the Committee may then approve its recommendation to the ACGIH® Board of Directors for adoption. If the Committee finds or receives substantive data that change its scientific opinion regarding an NIC TLV® or BEI®, the Committee may change its recommendation to the ACGIH® Board of Directors for the matter to be either retained on or withdrawn from the NIC. [Note: In the Physical Agents section of this book, the term Notice of Intent to Establish (NIE) is used in addition to NIC. For the purpose of this process overview, only the term NIC is used.]

- **TLV®/BEI® Policy Statement** — states what the TLVs® and BEIs® are and how they are intended to be used. While the TLVs® and BEIs® do contribute to the overall improvement in worker protection, the user must recognize the constraints and limitations subject to their proper use and bear the responsibility for such use (www.acgih.org/TLV/PolicyStmt.htm).

- **TLV®/BEI® Position Statement** — expresses ACGIH®'s position on the TLVs® and BEIs® process. ACGIH® is proud of the positive impact that the TLVs® and BEIs® have had on workers worldwide, and stands behind the hard work of its Committees to make the process more transparent and accessible. This section is presented in its entirety on pages v through vii (www.acgih.org/TLV/PosStmt.htm).

- **TLV®/BEI® Development Process** — gives an overview of the process the Committees go through when establishing a TLV® or BEI®. This section is presented in its entirety on pages viii through xiii (www.acgih.org/TLV/DevProcess.htm).

- **Committee Operations Manuals** — portable data files (PDF) of the Threshold Limit Values for Chemical Substances, the Threshold Limit Values for Physical Agents, and the Biological Exposure Indices Committees' Operations Manuals. Each Manual covers such areas as the Committee's mission, membership in the Committee, Committee make-up, internal and external communications with the Committee, flow of information, procedures for development of symposia and workshops, etc. (www.acgih.org/TLV/OpsManual.htm).

- **TLV®/BEI® Process Presentations** — stand-alone PowerPoint presentations from the annual American Industrial Hygiene Conference and Exposition (AIHce) are offered. These forums are open to all AIHce registrants and focus on the process used by ACGIH® and its TLV®, BEI®, and Bioaerosols Committees. These presentations are posted on the ACGIH® website (www.acgih.org/TLV/TLVPresentation.htm).

- **Under Study List** — contains substances, agents, and issues that are being considered by the Committees. Each Committee solicits data, comments, and suggestions that may assist in their deliberations about substances, agents, and issues on the Under Study list (www.acgih.org/TLV/Studies.htm). Further, each Committee solicits recommendations for additional chemical substances, physical agents, and issues of concern to the industrial hygiene and occupational health communities.

REVISIONS OR ADDITIONS
FOR 2010

All pertinent endnotes, abbreviations, and definitions relating to the materials in this publication appear on the inside back cover.

Chemical Substances Section

- Editorial revisions have been made to Appendix H.

- Proposed TLVs® that appeared on the 2009 NIC are adopted for the following substances:

Citral	α-Methyl styrene
Cotton dust, raw, untreated	Mineral oil, excluding
Cresol, all isomers	metal working fluids
Dieldrin	Portland cement
Hydrogen sulfide	Thallium and compounds, as Tl
Methyl isobutyl ketone	Thionyl chloride

- *Documentation* and adopted TLVs® are withdrawn for the following substances [*see also* Appendix G]:

Oil mist, mineral	Tantalum and Tantalum oxide dusts, as Ta

- The following chemical substances and proposed TLVs® new to this section are placed on the NIC:

Allyl bromide	Piperazine
2,4-Pentanedione	

- Revisions to adopted TLVs® are proposed for the following substances and placed on the NIC:

Acetic anhydride	Carbon black
Allyl chloride	Methylacrylonitrile
Calcium silicate	Methyl isopropyl ketone

- *Documentation* and adopted TLVs® for the following substances are proposed to be withdrawn:

Piperazine dihydrochloride (*see* NIC entry for Piperazine)	Soapstone (*see Documentation* for Talc)

- The following substances are retained on the NIC with revised TLV® recommendations or notations:

Ethyl benzene	4,4'-Thiobis(6-tert-butyl-m-cresol)
Maleic anhydride	

- Previously proposed TLVs® are retained on the NIC for the following substances:

Manganese, elemental and inorganic compounds as Mn	Toluene-2,4- or 2,6-diisocyanate

- *Documentation* were updated for the following without change to the recommended TLV®. *See* the 2010 Supplement to the *Documentation* of the TLVs® and BEIs®, 7th ed.:

Hexafluoropropylene	Talc
Silica, crystalline —	Wood dusts
α-quartz and cristobalite	

Biological Exposure Indices (BEIs®) Section

- Proposed BEIs® that appeared on the 2009 NIC are adopted for the following substances:

2-Methoxyethanol and	Toluene
2-Methoxyethyl acetate	Uranium
Methyl isobutyl ketone	

- *Documentation* and adopted BEI® for the following substance have been withdrawn due to new data:

  Vanadium pentoxide

- *Documentation* was updated for the following without change to the recommended BEI®. *See* the 2010 Supplement to the *Documentation* of the TLVs® and BEIs®, 7th ed.:

  Polycyclic aromatic hydrocarbons (PAHs)

- Negative Feasibility Assessment was determined and completed for the following:

Alachlor	Vanadium pentoxide

- There are no substances proposed for the 2010 Notice of Intended Changes.

Physical Agents Section

- The following agents that appeared on the 2009 NIC with revisions/additions are adopted:

  - SUB-RADIOFREQUENCY AND STATIC ELECTRIC FIELDS
  - RADIOFREQUENCY AND MICROWAVE RADIATION
  - ULTRAVIOLET RADIATION
  - IONIZING RADIATION

- The following agent is retained on the NIC with revisions/additions:

  - LASERS — The reason for this NIC is to add notes to Tables 2 and 3 "NTE" dual limits; to revise the TLVs® for pulse durations less than 50 ns and TLVs® between 1.4 and 1.5 µm; and to revise C_c.

Biologically Derived Airborne Contaminants Section

No new information for 2010.

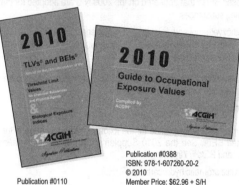

2010
Threshold Limit Values for Chemical Substances in the Work Environment

Adopted by ACGIH®
with Intended Changes

Contents

---

*Help ensure the continued development of
TLVs® and BEIs®. Make a tax deductible donation to
the FOHS Sustainable TLV®/BEI® Fund today!*

http://www.fohs.org/SusTLV-BEIPrgm.htm

INTRODUCTION TO THE CHEMICAL SUBSTANCES

General Information

The TLVs® are guidelines to be used by professional industrial hygienists. The values presented in this book are intended for use only as guidelines or recommendations to assist in the evaluation and control of potential workplace health hazards and for no other use (e.g., neither for evaluating or controlling community air pollution; nor for estimating the toxic potential of continuous, uninterrupted exposures or other extended work periods; nor for proving or disproving an existing disease or physical condition in an individual). Further, these values are not fine lines between safe and dangerous conditions and should not be used by anyone who is not trained in the discipline of industrial hygiene. TLVs® are not regulatory or consensus standards.

Editor's note: The approximate year that the current *Documentation* was last substantially reviewed and, where necessary, updated may be found following the CAS number for each of the adopted entries in the alphabetical listing, e.g., Aldrin [309-00-2] (2006). The reader is advised to refer to the "TLV® Chronology" section in each *Documentation* for a brief history of the TLV® recommendations and notations.

Definition of the TLVs®

Threshold Limit Values (TLVs®) refer to airborne concentrations of chemical substances and represent conditions under which it is believed that *nearly all* workers may be repeatedly exposed, day after day, over a working lifetime, without adverse health effects.

Those who use the TLVs® **MUST** consult the latest *Documentation* to ensure that they understand the basis for the TLV® and the information used in its development. The amount and quality of the information that is available for each chemical substance varies over time.

Chemical substances with equivalent TLVs® (i.e., same numerical values) cannot be assumed to have similar toxicologic effects or similar biologic potency. In this book, there are columns listing the TLVs® for each chemical substance (that is, airborne concentrations in parts per million [ppm] or milligrams per cubic meter [mg/m³]) and critical effects produced by the chemical substance. These critical effects form the basis of the TLV®.

ACGIH® recognizes that there will be considerable variation in the level of biological response to a particular chemical substance, regardless of the airborne concentration. Indeed, TLVs® do not represent a fine line between a healthy versus an unhealthy work environment or the point at which material impairment of health will occur. TLVs® will not adequately protect all workers. Some individuals may experience discomfort or even more serious adverse health effects when exposed to a chemical substance at the TLV® or even at concentrations below the TLV®. There are numerous possible reasons for increased susceptibility to a chemical substance, including age, gender, ethnicity, genetic factors (predisposition), lifestyle choices (e.g., diet, smoking, abuse of alcohol and other drugs), medications, and pre-existing medical conditions (e.g., aggravation of asthma or cardiovascular disease). Some individu-

TLV®-CS

als may become more responsive to one or more chemical substances following previous exposures (e.g., sensitized workers). Susceptibility to the effects of chemical substances may be altered during different periods of fetal development and throughout an individual's reproductive lifetime. Some changes in susceptibility may also occur at different work levels (e.g., light versus heavy work) or at exercise — situations in which there is increased cardiopulmonary demand. Additionally, variations in temperature (e.g., extreme heat or cold) and relative humidity may alter an individual's response to a toxicant. The *Documentation* for any given TLV® must be reviewed, keeping in mind that other factors may modify biological responses.

Although TLVs® refer to airborne levels of chemical exposure, dermal exposures may possibly occur in the workplace (*see* "Skin" on page 70 of the *Definitions and Notations* section).

Three categories of TLVs® are specified: time-weighted average (TWA); short-term exposure limit (STEL); and a ceiling (C). For most substances, a TWA alone or with a STEL is relevant. For some substances (e.g., irritant gases), only the TLV–C is applicable. If any of these TLV® types are exceeded, a potential hazard from that substance is presumed to exist.

Threshold Limit Value–Time-Weighted Average (TLV–TWA): The TWA concentration for a conventional 8-hour workday and a 40-hour workweek, to which it is believed that nearly all workers may be repeatedly exposed, day after day, for a working lifetime without adverse effect. Although calculating the average concentration for a workweek, rather than a workday, may be appropriate in some instances, ACGIH® does not offer guidance regarding such exposures.

Threshold Limit Value–Short-Term Exposure Limit (TLV–STEL): A 15-minute TWA exposure that should not be exceeded at any time during a workday, even if the 8-hour TWA is within the TLV–TWA. The TLV–STEL is the concentration to which it is believed that workers can be exposed continuously for a short period of time without suffering from 1) irritation, 2) chronic or irreversible tissue damage, 3) dose-rate-dependent toxic effects, or 4) narcosis of sufficient degree to increase the likelihood of accidental injury, impaired self-rescue, or materially reduced work efficiency. The TLV–STEL will not necessarily protect against these effects if the daily TLV–TWA is exceeded. The TLV–STEL usually supplements the TLV–TWA where there are recognized acute effects from a substance whose toxic effects are primarily of a chronic nature; however, the TLV–STEL may be a separate, independent exposure guideline. Exposures above the TLV–TWA up to the TLV–STEL should be less than 15 minutes, should occur no more than four times per day, and there should be at least 60 minutes between successive exposures in this range. An averaging period other than 15 minutes may be recommended when this is warranted by observed biological effects.

Threshold Limit Value–Ceiling (TLV–C): The concentration that should not be exceeded during any part of the working exposure. If instantaneous measurements are not available, sampling should be conducted for the minimum period of time sufficient to detect exposures at or above the ceiling value. ACGIH® believes that TLVs® based on physical irritation should be considered no less binding than those based on physical impairment. There is increasing evidence that physical irritation may initiate, promote, or accelerate adverse health effects through interaction with other chemical or biologic agents or through other mechanisms.

Excursion Limits

For many substances with a TLV–TWA, there is no TLV–STEL. Nevertheless, excursions above the TLV–TWA should be controlled, even where the 8-hour TLV–TWA is within recommended limits. Excursion limits apply to those TLV–TWAs that do not have TLV–STELs.

> Excursions in worker exposure levels may exceed 3 times the TLV–TWA for no more than a total of 30 minutes during a work-day, and under no circumstances should they exceed 5 times the TLV–TWA, provided that the TLV–TWA is not exceeded.

The approach here is that the maximum recommended excursion should be related to the variability generally observed in actual industrial processes. In reviewing large numbers of industrial hygiene surveys conducted by the U.S. National Institute for Occupational Safety and Health, Leidel et al. (1975) found that short-term exposure measurements were generally lognormally distributed.

While a complete discussion of the theory and properties of the lognormal distribution is beyond the scope of this section, a brief description of some important terms is presented. The measure of central tendency in a lognormal distribution is the antilog of the mean logarithm of the sample values. The distribution is skewed, and the geometric mean (m_g) is always smaller than the arithmetic mean by an amount that depends on the geometric standard deviation. In the lognormal distribution, the geometric standard deviation (sd_g) is the antilog of the standard deviation of the sample value logarithms, and 68.26% of all values lie between m_g/sd_g and $m_g \times sd_g$.

If the short-term exposure values in a given situation have a geometric standard deviation of 2.0, 5% of all values will exceed 3.13 times the geometric mean. If a process displays variability greater than this, it is not under good control, and efforts should be made to restore control.

The approach is a considerable simplification of the lognormal concentration distribution concept but is considered more convenient. If exposure excursions are maintained within the recommended limits, the geometric standard deviation of the concentration measurements will be near 2.0, and the goal of the recommendations will be accomplished. It is recognized that the geometric standard deviations of some common workplace exposures may exceed 2.0 (Buringh and Lanting, 1991). If such distributions are known and workers are not at increased risk of adverse health effects, recommended excursion limits should be modified based upon workplace-specific data. When the toxicologic data for a specific substance are available to establish a TLV–STEL or a TLV–C, these values take precedence over the excursion limit.

TWA and STEL versus Ceiling (C)

A substance may have certain toxicological properties that require the use of a TLV–C rather than a TLV–TWA excursion limit or a TLV–STEL. The amount by which the TLVs® may be exceeded for short periods without injury to health depends upon a number of factors such as the nature of the contaminant, whether very high concentrations — even for short periods — produce acute poisoning, whether the effects are cumulative, the frequency with which high concentrations occur, and the duration of such periods. All factors must be taken into consideration in arriving at a decision as to whether a hazardous condition exists.

Although the TWA concentration provides the most satisfactory, practical way of monitoring airborne agents for compliance with the TLVs®, there are certain substances for which it is inappropriate. In the latter group are substances that are predominantly fast-acting and whose TLV® is more appropriately based on this particular response. Substances with this type of response are best controlled by a TLV–C that should not be exceeded. It is implicit in these definitions that the manner of sampling to determine noncompliance with the TLVs® for each group must differ. Consequently, a single, brief sample that is applicable to a TLV–C is not appropriate to the TLV–TWA; here, a sufficient number of samples are needed to permit determination of a TWA concentration throughout a complete cycle of operation or throughout the workshift.

Whereas the TLV–C places a definite boundary that exposure concentrations should not be permitted to exceed, the TLV–TWA requires an explicit limit to the excursions which are acceptable above the recommended TLV–TWAs.

Mixtures

Special consideration should also be given to the application of the TLVs® in assessing the health hazards that may be associated with exposure to a mixture of two or more substances. A brief discussion of basic considerations involved in developing TLVs® for mixtures and methods for their development, amplified by specific examples, is given in Appendix E.

Deviations in Work Conditions and Work Schedules

Application of TLVs® to Unusual Ambient Conditions

When workers are exposed to air contaminants at temperatures and pressures substantially different than those at normal temperature and pressure (NTP) conditions (25°C and 760 torr), care should be taken in comparing sampling results to the applicable TLVs®. For aerosols, the TWA exposure concentration (calculated using sample volumes not adjusted to NTP conditions) should be compared directly to the applicable TLVs® published in the *TLVs® and BEIs®* book. For gases and vapors, there are a number of options for comparing air-sampling results to the TLV®, and these are discussed in detail by Stephenson and Lillquist (2001). One method that is simple in its conceptual approach is 1) to determine the exposure concentration, expressed in terms of mass per volume, at the sampling site using the sample volume not adjusted to NTP conditions, 2) if required, to convert the TLV® to mg/m³ (or other mass per volume measure) using a molar volume of 24.4 L/mole, and 3) to compare the exposure concentration to the TLV®, both in units of mass per volume.

A number of assumptions are made when comparing sampling results obtained under unusual atmospheric conditions to the TLVs®. One such assumption is that the volume of air inspired by the worker per workday is not appreciably different under moderate conditions of temperature and pressure as compared to NTP (Stephenson and Lillquist, 2001). An additional assumption for gases and vapors is that absorbed dose is correlated to the partial pressure of the inhaled compound. Sampling results obtained under unusual conditions cannot easily be compared to the published TLVs®, and extreme care should be exercised if workers are exposed to very high or low ambient pressures.

Unusual Work Schedules

Application of TLVs® to work schedules markedly different from the conventional 8-hour day, 40-hour workweek requires particular judgment to provide protection for these workers equal to that provided to workers on conventional work shifts. Short workweeks can allow workers to have more than one job, perhaps with similar exposures, and may result in overexposure, even if neither job by itself entails overexposure.

Numerous mathematical models to adjust for unusual work schedules have been described. In terms of toxicologic principles, their general objective is to identify a dose that ensures that the daily peak body burden or weekly peak body burden does not exceed that which occurs during a normal 8-hour/day, 5-day/week shift. A comprehensive review of the approaches to adjusting occupational exposure limits for unusual work schedules is provided in *Patty's Industrial Hygiene* (Paustenbach, 2000). Other selected readings on this topic include Lapare et al. (2003), Brodeur et al. (2001), Caldwell et al. (2001), Eide (2000), Verma (2000), Roach (1978), and Hickey and Reist (1977).

Another model that addresses unusual work schedules is the Brief and Scala model (1986), which is explained in detail in *Patty's Industrial Hygiene* (Paustenbach, 2000). This model reduces the TLV® proportionately for both increased exposure time and reduced recovery (i.e., non-exposure) time, and is generally intended to apply to work schedules longer than 8 hours/day or 40 hours/week. The model should not be used to justify very high exposures as "allowable" where the exposure periods are short (e.g., exposure to 8 times the TLV–TWA for 1 hour and zero exposure during the remainder of the shift). In this respect, the general limitations on TLV–TWA excursions and TLV–STELs should be applied to avoid inappropriate use of the model with very short exposure periods or shifts.

The Brief and Scala model is easier to use than some of the more complex models based on pharmacokinetic actions. The application of such models usually requires knowledge of the biological half-life of each substance, and some models require additional data. Another model developed by the University of Montreal and the Institute de Recherche en Sante et en Securite du Travail (IRSST) uses the Haber method to calculate adjusted exposure limits (Brodeur et al., 2001). This method generates values close to those obtained from physiologically based pharmacokinetic (PBPK) models.

Because adjusted TLVs® do not have the benefit of historical use and long-time observation, medical supervision during initial use of adjusted TLVs® is advised. Unnecessary exposure of workers should be avoided, even if a model shows such exposures to be "allowable." Mathematical models should not be used to justify higher-than-necessary exposures.

TLV® Units

TLVs® are expressed in ppm or mg/m³. An inhaled chemical substance may exist as a gas, vapor, or aerosol.

- A gas is a chemical substance whose molecules are moving freely within a space in which they are confined (e.g., cylinder/tank) at normal temperature and pressure (NTP). Gases assume no shape or volume.
- A vapor is the gaseous phase of a chemical substance that exists as a liquid or a solid at NTP. The amount of vapor given off by a chemical substance is expressed as the vapor pressure and is a function of temperature and pressure.
- An aerosol is a suspension of solid particles or liquid droplets in a gaseous medium. Other terms used to describe an aerosol include dust, mist, fume, fog, fiber, smoke, and smog. Aerosols may be characterized by their aerodynamic behavior and the site(s) of deposition in the human respiratory tract.

TLVs® for aerosols are usually established in terms of mass of the chemical substance in air by volume. These TLVs® are expressed in mg/m³.

TLVs® for gases and vapors are established in terms of parts of vapor or gas per million parts of contaminated air by volume (ppm), but may also be expressed in mg/m³. For convenience to the user, these TLVs® also reference molecular weights. Where 24.45 = molar volume of air in liters at NTP conditions (25°C and 760 torr), the conversion equations for gases and vapors [ppm ↔ mg/m³] are as follows:

$$\text{TLV in ppm} = \frac{(\text{TLV in mg/m}^3)\,(24.45)}{(\text{gram molecular weight of substance})}$$

OR

$$\text{TLV in mg/m}^3 = \frac{(\text{TLV in ppm})\,(\text{gram molecular weight of substance})}{24.45}$$

When converting values expressed as an element (e.g., as Fe, as Ni), the molecular weight of the element should be used, not that of the entire compound.

In making conversions for substances with variable molecular weights, appropriate molecular weights should be estimated or assumed (*see* the TLV® *Documentation*).

User Information

Each TLV® is supported by a comprehensive *Documentation*. It is imperative to consult the latest *Documentation* when applying the TLV®.

Additional copies of the *TLVs® and BEIs®* book and the multi-volume *Documentation of the Threshold Limit Values and Biological Exposure Indices*, upon which this book is based, are available from ACGIH®. *Documentation* of individual TLVs® is also available. Consult the ACGIH® website (www.acgih.org/store) for additional information and availability concerning these publications.

References and Selected Readings

Brief RS; Scala RA: Occupational health aspects of unusual work schedules: a review of Exxon's experiences. Am Ind Hyg Assoc J 47(4):199-202 (1986).

Brodeur J; Vyskocil A; Tardif R; et al.: Adjustment of permissible exposure values to unusual work schedules. Am Ind Hyg Assoc J 62:584-594 (2001).

Buringh E; Lanting R: Exposure variability in the workplace: its implications for the assessment of compliance. Am Ind Hyg Assoc J 52:6-13 (1991).

Caldwell DJ; Armstrong TW; Barone NJ; et al.: Lessons learned while compiling a quantitative exposure database from the published literature. Appl Occup Environ Hyg 16(2):174-177 (2001).

Eide I: The application of 8-hour occupational exposure limits to non-standard work schedules offshore. Ann Occup Hyg 34(1):13-17 (1990).

Hickey JL; Reist PC: Application of occupational exposure limits to unusual work schedules. Am Ind Hyg Assoc J 38(11):613-621 (1977).

Lapare S; Brodeur J; Tardif R: Contribution of toxicokinetic modeling to the adjustment of exposure limits to unusual work schedules. Am Ind Hyg Assoc J 64(1):17-23 (2003).

Leidel NA; Busch KA; Crouse WE: Exposure measurement action level and occupational environmental variability. DHEW (NIOSH) Pub. No. 76-131; NTIS Pub. No. PB- 267-509. U.S. National Technical Information Service, Springfield, VA (December 1975).

Paustenbach DJ: Pharmacokinetics and Unusual Work Schedules. In: Patty's Industrial Hygiene, 5th ed., Vol. 3, Part VI, Law, Regulation, and Management, Chap. 40, pp. 1787-1901. RL Harris, Ed. John Wiley & Sons, Inc., New York (2000).

Roach SA: Threshold limit values for extraordinary work schedules. Am Ind Hyg Assoc J 39(4):345-348 (1978).

Stephenson DJ; Lillquist DR: The effects of temperature and pressure on airborne exposure concentrations when performing compliance evaluations using ACGIH TLVs and OSHA PELs. Appl Occup Environ Hyg 16(4):482-486 (2001).

Verma DK: Adjustment of occupational exposure limits for unusual work schedules. Am Ind Hyg Assoc J 61(3):367-374 (2000).

TLV®–CS

All pertinent notes relating to the material in the Chemical Substances section of this book appear in the appendices for this section or on the inside back cover.

TLV®-CS

Substance [CAS No.] (Documentation date)	ADOPTED VALUES					
	TWA	STEL	Notations	MW	TLV® Basis	
Acetaldehyde [75-07-0] (1992)	—	C 25 ppm	A3	44.05	Eye & URT irr	
Acetic acid [64-19-7] (2003)	10 ppm	15 ppm	—	60.00	URT & eye irr; pulm func	
‡ Acetic anhydride [108-24-7] (1990)	(5 ppm)	(—)	(—)	102.09	(Eye & URT irr)	
Acetone [67-64-1] (1996)	500 ppm	750 ppm	A4; BEI	58.05	URT & eye irr; CNS impair; hematologic eff	
Acetone cyanohydrin [75-86-5], as CN (1991)	—	C 5 mg/m³	Skin	85.10	URT irr; headache; hypoxia/cyanosis	
Acetonitrile [75-05-8] (1996)	20 ppm	—	Skin; A4	41.05	LRT irr	
Acetophenone [98-86-2] (2008)	10 ppm	—	—	120.15	Eye irr	
Acetylene [74-86-2] (1990)	Simple asphyxiant [D]			26.02	Asphyxia	
Acetylsalicylic acid (Aspirin) [50-78-2] (1977)	5 mg/m³	—	—	180.15	Skin & eye irr	
Acrolein [107-02-8] (1995)	—	C 0.1 ppm	Skin; A4	56.06	Eye & URT irr; pulm edema; pulm emphysema	
Acrylamide [79-06-1] (2004)	0.03 mg/m³ [IFV]	—	Skin; A3	71.08	CNS impair	
Acrylic acid [79-10-7] (1986)	2 ppm	—	Skin; A4	72.06	URT irr	
Acrylonitrile [107-13-1] (1997)	2 ppm	—	Skin; A3	53.05	CNS impair; LRT irr	
Adipic acid [124-04-9] (1990)	5 mg/m³	—	—	146.14	URT irr; ANS impair	

Substance [CAS No.] (*Documentation date*)	ADOPTED VALUES			MW	TLV® Basis
	TWA	STEL	Notations		
Adiponitrile [111-69-3] (1990)	2 ppm	—	Skin	108.10	URT & LRT irr
Alachlor [15972-60-8] (2006)	1 mg/m³ (IFV)	—	SEN; A3	269.8	Hemosiderosis
Aldrin [309-00-2] (2006)	0.05 mg/m³ (IFV)	—	Skin; A3	364.93	CNS impair; liver & kidney dam
Aliphatic hydrocarbon gases Alkanes [C₁–C₄] (2001)	1000 ppm	—	—	Varies	Card sens; CNS impair
Allyl alcohol [107-18-6] (1996)	0.5 ppm	—	Skin; A4	58.08	Eye & URT irr
‡ Allyl chloride [107-05-1] (1990)	1 ppm	2 ppm	( ); A3	76.50	Eye & URT irr; liver & kidney dam
Allyl glycidyl ether (AGE) [106-92-3] (1995)	1 ppm	—	A4	114.14	URT irr; dermatitis; eye & skin irr
Allyl propyl disulfide [2179-59-1] (2001)	0.5 ppm	—	SEN	148.16	URT & eye irr
Aluminum metal [7429-90-5] and insoluble compounds (2007)	1 mg/m³ (R)	—	A4	26.98 Varies	Pneumoconiosis; LRT irr; neurotoxicity
4-Aminodiphenyl [92-67-1] (1968)	—(L)	—	Skin; A1	169.23	Bladder & liver cancer
2-Aminopyridine [504-29-0] (1966)	0.5 ppm	—	—	91.11	Headache; nausea; CNS impair; dizziness
Amitrole [61-82-5] (1983)	0.2 mg/m³	—	A3	84.08	Thyroid eff
Ammonia [7664-41-7] (1970)	25 ppm	35 ppm	—	17.03	Eye dam; URT irr

TLV®–CS

Substance [CAS No.] (*Documentation date*)	ADOPTED VALUES			MW	TLV® Basis
	TWA	STEL	Notations		
Ammonium chloride fume [12125-02-9] (1970)	10 mg/m³	20 mg/m³	—	53.50	Eye & URT irr
Ammonium perfluorooctanoate [3825-26-1] (1992)	0.01 mg/m³	—	Skin; A3	431.00	Liver dam
Ammonium sulfamate [7773-06-0] (1956)	10 mg/m³	—	—	114.13	—
tert-Amyl methyl ether (TAME) [994-05-8] (1999)	20 ppm	—	—	102.2	CNS impair; Embryo/fetal dam
Aniline [62-53-3] (1979)	2 ppm	—	Skin; A3; BEI	93.12	MeHb-emia
o-Anisidine [90-04-0] (1979)	0.5 mg/m³	—	Skin; A3; BEI$_M$	123.15	MeHb-emia
p-Anisidine [104-94-9] (1979)	0.5 mg/m³	—	Skin; A4; BEI$_M$	123.15	MeHb-emia
Antimony [7440-36-0] and compounds, as Sb (1979)	0.5 mg/m³	—	—	121.75	Skin & URT irr
Antimony hydride [7803-52-3] (1990)	0.1 ppm	—	—	124.78	Hemolysis; kidney dam; LRT irr
Antimony trioxide [1309-64-4] production (1977)	—(L)	—	A2	291.5	Lung cancer; pneumoconiosis
ANTU [86-88-4] (1990)	0.3 mg/m³	—	A4; Skin	202.27	Thyroid eff; nausea
Argon [7440-37-1] (1990)	Simple asphyxiant (D)	Simple asphyxiant (D)		39.95	Asphyxia
Arsenic [7440-38-2] and inorganic compounds, as As (1990)	0.01 mg/m³	—	A1; BEI	74.92 Varies	Lung cancer
Arsine [7784-42-1] (2006)	0.005 ppm	—	—	77.95	PNS & vascular system impair; kidney & liver impair

Substance [CAS No.] (*Documentation* date)	ADOPTED VALUES			Notations	MW	TLV® Basis
	TWA	STEL				
Asbestos, all forms [1332-21-4] (1994)	0.1 f/cc [F]	—		A1	NA	Pneumoconiosis; lung cancer; mesothelioma
Asphalt (Bitumen) fume [8052-42-4], as benzene-soluble aerosol (1999)	0.5 mg/m³ [I]	—		A4; BEI_P	—	URT & eye irr
Atrazine [1912-24-9] (1985)	5 mg/m³	—		A4	216.06	CNS convul
Azinphos-methyl [86-50-0] (1999)	0.2 mg/m³ (IFV)	—		Skin; SEN; A4; BEI_A	317.34	Cholinesterase inhib
Barium [7440-39-3] and soluble compounds, as Ba (1990)	0.5 mg/m³	—		A4	137.30	Eye, skin, & GI irr; muscular stim
Barium sulfate [7727-43-7] (1983)	10 mg/m³	—		—	233.43	Pneumoconiosis
Benomyl [17804-35-2] (2007)	1 mg/m³ [I]	—		SEN; A3	290.32	URT irr; male repro & testicular dam; embryo/fetal dam
Benz[a]anthracene [56-55-3] (1990)	— [L]	—		A2; BEI_P	228.30	Skin cancer
Benzene [71-43-2] (1996)	0.5 ppm	2.5 ppm		Skin; A1; BEI	78.11	Leukemia
Benzidine [92-87-5] (1979)	— [L]	—		Skin; A1	184.23	Bladder cancer
Benzo[b]fluoranthene [205-99-2] (1990)	— [L]	—		A2; BEI_P	252.30	Cancer
Benzo[a]pyrene [50-32-8] (1990)	— [L]	—		A2; BEI_P	252.30	Cancer

Substance [CAS No.] (Documentation date)	ADOPTED VALUES TWA	STEL	Notations	MW	TLV® Basis
Benzotrichloride [98-07-7] (1994)	—	C 0.1 ppm	Skin; A2	195.50	Eye, skin, & URT irr
Benzoyl chloride [98-88-4] (1992)	—	C 0.5 ppm	A4	140.57	URT & eye irr
Benzoyl peroxide [94-36-0] (1990)	5 mg/m³	—	A4	242.22	URT & skin irr
Benzyl acetate [140-11-4] (1990)	10 ppm	—	A4	150.18	URT irr
Benzyl chloride [100-44-7] (1990)	1 ppm	—	A3	126.58	Eye, skin, & URT irr
Beryllium [7440-41-7] and compounds, as Be (2008)	0.00005 mg/m³ (I)	—	Skin; SEN; A1	9.01	Beryllium sens; chronic beryllium disease (berylliosis)
Biphenyl [92-52-4] (1979)	0.2 ppm	—	—	154.20	Pulm func
Bis (2-dimethylaminoethyl) ether (DMAEE) [3033-62-3] (1997)	0.05 ppm	0.15 ppm	Skin	160.26	URT, eye, & skin irr
Bismuth telluride (1970) Undoped [1304-82-1] Se-doped, as Bi₂Te₃	10 mg/m³ 5 mg/m³	— —	A4 A4	800.83	Lung dam
Borate compounds, inorganic [1330-43-4; 1303-96-4; 10043-35-3; 12179-04-3] (2004)	2 mg/m³ (I)	6 mg/m³ (I)	A4	Varies	URT irr
Boron oxide [1303-86-2] (1985)	10 mg/m³	—	—	69.64	Eye & URT irr

| Substance [CAS No.] (*Documentation date*) | ADOPTED VALUES | | | | TLV® Basis |
	TWA	STEL	Notations	MW	
Boron tribromide [10294-33-4] (1990)	—	C 1 ppm	—	250.57	URT irr
Boron trifluoride [7637-07-2] (1962)	—	C 1 ppm	—	67.82	LRT irr; pneumonitis
Bromacil [314-40-9] (1976)	10 mg/m^3	—	A3	261.11	Thyroid eff
Bromine [7726-95-6] (1991)	0.1 ppm	0.2 ppm	—	159.81	URT & LRT irr; lung dam
Bromine pentafluoride [7789-30-2] (1979)	0.1 ppm	—	—	174.92	Eye, skin, & URT irr
Bromoform [75-25-2] (2008)	0.5 ppm	—	A3	252.73	Liver dam; URT & eye irr
1-Bromopropane [106-94-5] (2003)	10 ppm	—	—	122.99	Liver & embryo/fetal dam; neurotoxicity
1,3-Butadiene [106-99-0] (1994)	2 ppm	—	A2	54.09	Cancer
Butane, all isomers [106-97-8; 75-28-5]	See Aliphatic hydrocarbon gases: Alkanes [C$_1$–C$_4$]				
n-Butanol [71-36-3] (1998)	20 ppm	—	—	74.12	Eye & URT irr
sec-Butanol [78-92-2] (2001)	100 ppm	—	—	74.12	URT irr; CNS impair
tert-Butanol [75-65-0] (1992)	100 ppm	—	A4	74.12	CNS impair
Butenes, all isomers [106-98-9; 107-01-7; 590-18-1; 624-64-6; 25167-67-3]	250 ppm	—	—	56.11	Body weight eff
Isobutene [115-11-7] (2007)	250 ppm	—	A4	—	URT irr; body weight eff

Substance [CAS No.] (*Documentation* date)	ADOPTED VALUES					
	TWA	STEL	Notations	MW	TLV® Basis	
2-Butoxyethanol (EGBE) [111-76-2] (1996)	20 ppm	—	A3	118.17	Eye & URT irr	
2-Butoxyethyl acetate (EGBEA) [112-07-2] (2000)	20 ppm	—	A3	160.2	Hemolysis	
n-Butyl acetate [123-86-4] (1995)	150 ppm	200 ppm	—	116.16	Eye & URT irr	
sec-Butyl acetate [105-46-4] (1965)	200 ppm	—	—	116.16	Eye & URT irr	
tert-Butyl acetate [540-88-5] (1965)	200 ppm	—	—	116.16	Eye & URT irr	
n-Butyl acrylate [141-32-2] (1996)	2 ppm	—	SEN; A4	128.17	Skin, eye, & URT irr	
n-Butylamine [109-73-9] (1985)	—	C 5 ppm	Skin	73.14	Headache; URT & eye irr	
Butylated hydroxytoluene (BHT) [128-37-0] (2001)	2 mg/m^3 (IFV)	—	A4	220.34	URT irr	
tert-Butyl chromate, as CrO$_3$ [1189-85-1] (1960)	—	C 0.1 mg/m^3	Skin	230.22	LRT & skin irr	
n-Butyl glycidyl ether (BGE) [2426-08-6] (2002)	3 ppm	—	Skin; SEN	130.21	Repro dam	
n-Butyl lactate [138-22-7] (1973)	5 ppm	—	—	146.19	Headache; URT irr	
n-Butyl mercaptan [109-79-5] (1968)	0.5 ppm	—	—	90.19	URT irr	
o-sec-Butylphenol [89-72-5] (1977)	5 ppm	—	Skin	150.22	URT, eye, & skin irr	
p-tert-Butyl toluene [98-51-1] (1990)	1 ppm	—	—	148.18	Eye & URT irr; nausea	

	ADOPTED VALUES				
Substance [CAS No.] (*Documentation date*)	TWA	STEL	Notations	MW	TLV® Basis
Cadmium [7440-43-9] and compounds, as Cd (1990)	0.01 mg/m³ 0.002 mg/m³ (R)	— —	A2; BEI A2; BEI	112.40 Varies	Kidney dam
Calcium chromate [13765-19-0], as Cr (1988)	0.001 mg/m³	—	A2	156.09	Lung cancer
Calcium cyanamide [156-62-7] (1973)	0.5 mg/m³	—	A4	80.11	Eye & URT irr
Calcium hydroxide [1305-62-0] (1979)	5 mg/m³	—	—	74.10	Eye, URT, & skin irr
Calcium oxide [1305-78-8] (1990)	2 mg/m³	—	—	56.08	URT irr
‡ (Calcium silicate, synthetic nonfibrous) [1344-95-2] (1988)	(10 mg/m³ (E))	—	A4	—	(URT irr)
Calcium sulfate [7778-18-9; 10034-76-1; 10101-41-4; 13397-24-5] (2005)	10 mg/m³ (I)	—	—	136.14	Nasal symptoms
Camphor, synthetic [76-22-2] (1990)	2 ppm	3 ppm	A4	152.23	Eye & URT irr, anosmia
Caprolactam [105-60-2] (1997)	5 mg/m³ (IFV)	—	A5	113.16	URT irr
Captafol [2425-06-1] (1990)	0.1 mg/m³	—	Skin; A4	349.06	Skin irr
Captan [133-06-2] (1999)	5 mg/m³ (I)	—	SEN; A3	300.60	Skin irr
Carbaryl [63-25-2] (2007)	0.5 mg/m³ (IFV)	—	Skin; A4; BEI_A	201.20	Cholinesterase inhib; male repro dam; embryo dam

| Substance [CAS No.] (Documentation date) | ADOPTED VALUES | | | | |
	TWA	STEL	Notations	MW	TLV® Basis
Carbofuran [1563-66-2] (2001)	0.1 mg/m³ (IFV)	—	A4; BEI_A	221.30	Cholinesterase inhib
‡ Carbon black [1333-86-4] (1985)	(3.5 mg/m³)	—	(A4)	—	( )
Carbon dioxide [124-38-9] (1983)	5000 ppm	30,000 ppm	—	44.01	Asphyxia
Carbon disulfide [75-15-0] (2005)	1 ppm	—	Skin; A4; BEI	76.14	PNS impair
Carbon monoxide [630-08-0] (1989)	25 ppm	—	BEI	28.01	COHb-emia
Carbon tetrabromide [558-13-4] (1972)	0.1 ppm	0.3 ppm	—	331.65	Liver dam; eye, URT, & skin irr
Carbon tetrachloride [56-23-5] (1990)	5 ppm	10 ppm	Skin; A2	153.84	Liver dam
Carbonyl fluoride [353-50-4] (1990)	2 ppm	5 ppm	—	66.01	LRT irr; bone dam
Catechol [120-80-9] (1985)	5 ppm	—	Skin; A3	110.11	Eye & URT irr; dermatitis
Cellulose [9004-34-6] (1985)	10 mg/m³	—	—	NA	URT irr
Cesium hydroxide [21351-79-1] (1990)	2 mg/m³	—	—	149.92	URT, skin, & eye irr
Chlordane [57-74-9] (1985)	0.5 mg/m³	—	Skin; A3	409.80	Liver dam
Chlorinated camphene [8001-35-2] (1990)	0.5 mg/m³	1 mg/m³	Skin; A3	414.00	CNS convul; liver dam
o-Chlorinated diphenyl oxide [31242-93-0] (1979)	0.5 mg/m³	—	—	377.00	Chloracne; liver dam
Chlorine [7782-50-5] (1986)	0.5 ppm	1 ppm	A4	70.91	URT & eye irr

TLV®–CS

| Substance [CAS No.] *(Documentation date)* | ADOPTED VALUES | | | | |
	TWA	STEL	Notations	MW	TLV® Basis
Chlorine dioxide [10049-04-4] (1991)	0.1 ppm	0.3 ppm	—	67.46	LRT irr; bronchitis
Chlorine trifluoride [7790-91-2] (1979)	—	C 0.1 ppm	—	92.46	Eye & URT irr; lung dam
Chloroacetaldehyde [107-20-0] (1990)	—	C 1 ppm	—	78.50	URT & eye irr
Chloroacetone [78-95-5] (1986)	—	C 1 ppm	Skin	92.53	Eye & URT irr
2-Chloroacetophenone [532-27-4] (1990)	0.05 ppm	—	A4	154.59	Eye, URT, & skin irr
Chloroacetyl chloride [79-04-9] (1988)	0.05 ppm	0.15 ppm	Skin	112.95	URT irr
Chlorobenzene [108-90-7] (1988)	10 ppm	—	A3; BEI	112.56	Liver dam
o-Chlorobenzylidene malononitrile [2698-41-1] (1990)	—	C 0.05 ppm	Skin; A4	188.61	URT irr; skin sens
Chlorobromomethane [74-97-5] (2008)	200 ppm	—	—	129.39	CNS impair; liver dam
Chlorodifluoromethane [75-45-6] (1990)	1000 ppm	—	A4	86.47	CNS impair; asphyxia; card sens
Chlorodiphenyl (42% chlorine) [53469-21-9] (1979)	1 mg/m³	—	Skin	266.50	Liver dam; eye irr; chloracne
Chlorodiphenyl (54% chlorine) [11097-69-1] (1990)	0.5 mg/m³	—	Skin; A3	328.40	URT irr; liver dam; chloracne
Chloroform [67-66-3] (1990)	10 ppm	—	A3	119.38	Liver dam; embryo/fetal dam; CNS impair

TLV®–CS

Substance [CAS No.] (Documentation date)	ADOPTED VALUES					
	TWA	STEL	Notations	MW	TLV® Basis	
bis(Chloromethyl) ether [542-88-1] (1979)	0.001 ppm	—	A1	114.96	Lung cancer	
Chloromethyl methyl ether [107-30-2] (1979)	—(L)	—	A2	80.50	Lung cancer	
1-Chloro-1-nitropropane [600-25-9] (1971)	2 ppm	—	—	123.54	Eye irr; pulm edema	
Chloropentafluoroethane [76-15-3] (1978)	1000 ppm	—	—	154.47	Card sens	
Chloropicrin [76-06-2] (1990)	0.1 ppm	—	A4	164.39	Eye irr; pulm edema	
1-Chloro-2-propanol [127-00-4] and 2-Chloro-1-propanol [78-89-7] (1999)	1 ppm	—	Skin; A4	94.54	Liver dam	
β-Chloroprene [126-99-8] (1990)	10 ppm	—	Skin	88.54	URT & eye irr	
2-Chloropropionic acid [598-78-7] (1988)	0.1 ppm	—	Skin	108.53	Male repro dam	
o-Chlorostyrene [2039-87-4] (1972)	50 ppm	75 ppm	—	138.60	CNS impair; peripheral neuropathy	
o-Chlorotoluene [95-49-8] (1971)	50 ppm	—	—	126.59	URT, eye, & skin irr	
Chlorpyrifos [2921-88-2] (2000)	0.1 mg/m³ (IFV)	—	Skin; A4; BEI_A	350.57	Cholinesterase inhib	
Chromite ore processing (Chromate), as Cr	0.05 mg/m³	—	A1	—	Lung cancer	

Substance [CAS No.] (Documentation date)	ADOPTED VALUES			Notations	MW	TLV® Basis
	TWA	STEL				
Chromium, [7440-47-3] and inorganic compounds, as Cr (1991)						
Metal and Cr III compounds	0.5 mg/m³	—		A4	Varies	URT & skin irr
Water-soluble Cr VI compounds	0.05 mg/m³	—		A1; BEI	Varies	URT irr; cancer
Insoluble Cr VI compounds	0.01 mg/m³	—		A1	Varies	Lung cancer
Chromyl chloride [14977-61-8] (1990)	0.025 ppm	—		—	154.92	URT & skin irr
Chrysene [218-01-9] (1990)	— (L)	—		A3; BEI$_P$	228.30	Cancer
* Citral [5392-40-5] (2009)	5 ppm (IFV)	—		Skin; SEN; A4	152.24	Body weight eff; URT irr; eye dam
Clopidol [2971-90-6] (1972)	10 mg/m³	—		A4	192.06	URT irr
Coal dust (1995)						
Anthracite	0.4 mg/m³ (R)	—		A4	—	Lung dam; pulm fibrosis
Bituminous	0.9 mg/m³ (R)	—		A4	—	Lung dam; pulm fibrosis
Coal tar pitch volatiles [65996-93-2], as benzene soluble aerosol (1984)	0.2 mg/m³	—		A1; BEI$_P$	—	Cancer
Cobalt [7440-48-4] and inorganic compounds, as Co (1993)	0.02 mg/m³	—		A3; BEI	58.93 Varies	Asthma; pulm func; myocardial eff
Cobalt carbonyl [10210-68-1], as Co (1980)	0.1 mg/m³	—		—	341.94	Pulm edema; spleen dam

TLV®-CS

Substance [CAS No.] (*Documentation date*)	ADOPTED VALUES				
	TWA	STEL	Notations	MW	TLV® Basis
Cobalt hydrocarbonyl [16842-03-8], as Co (1980)	0.1 mg/m³	—	—	171.98	Pulm edema; lung dam
Copper [7440-50-8] (1990)				63.55	Irr; GI; metal fume fever
Fume	0.2 mg/m³	—	—		
Dusts and mists, as Cu	1 mg/m³	—	—		
* Cotton dust, raw, untreated (2009)	0.1 mg/m³ (T)	—	A4	—	Byssinosis; bronchitis; pulm func
Coumaphos [56-72-4] (2005)	0.05 mg/m³ (IFV)	—	Skin; A4; BEI_A	362.8	Cholinesterase inhib
* Cresol, all isomers (2009) [1319-77-3; 95-48-7; 108-39-4; 106-44-5]	20 mg/m³ (IFV)	—	Skin; A4	108.14	URT irr
Crotonaldehyde [4170-30-3] (1995)	—	C 0.3 ppm	Skin; A3	70.09	Eye & URT irr
Crufomate [299-86-5] (1971)	5 mg/m³	—	A4; BEI_A	291.71	Cholinesterase inhib
Cumene [98-82-8] (1997)	50 ppm	—	—	120.19	Eye, skin, & URT irr; CNS impair
Cyanamide [420-04-2] (1974)	2 mg/m³	—	—	42.04	Skin & eye irr
Cyanogen [460-19-5] (1966)	10 ppm	—	—	52.04	LRT & eye irr
Cyanogen chloride [506-77-4] (1977)	—	C 0.3 ppm	—	61.48	Pulm edema; eye, skin, & URT irr
Cyclohexane [110-82-7] (1964)	100 ppm	—	—	84.16	CNS impair
Cyclohexanol [108-93-0] (1979)	50 ppm	—	Skin	100.16	Eye irr; CNS impair

Substance [CAS No.] (Documentation date)	ADOPTED VALUES				
	TWA	STEL	Notations	MW	TLV® Basis
Cyclohexanone [108-94-1] (1990)	20 ppm	50 ppm	Skin; A3	98.14	Eye & URT irr
Cyclohexene [110-83-8] (1964)	300 ppm	—	—	82.14	URT & eye irr
Cyclohexylamine [108-91-8] (1990)	10 ppm	—	A4	99.17	URT & eye irr
Cyclonite [121-82-4] (1994)	0.5 mg/m³	—	Skin; A4	222.26	Liver dam
Cyclopentadiene [542-92-7] (1963)	75 ppm	—	—	66.10	URT & eye irr
Cyclopentane [287-92-3] (1978)	600 ppm	—	—	70.13	URT, eye, & skin irr; CNS impair
Cyhexatin [13121-70-5] (1990)	5 mg/m³	—	A4	385.16	URT irr; body weight eff; kidney dam
2,4-D [94-75-7] (1990)	10 mg/m³	—	A4	221.04	URT & skin irr
DDT [50-29-3] (1979)	1 mg/m³	—	A3	354.50	Liver dam
Decaborane [17702-41-9] (1979)	0.05 ppm	0.15 ppm	Skin	122.31	CNS convul; cognitive decrement
Demeton [8065-48-3] (1998)	0.05 mg/m³ (IFV)	—	Skin; BEIA	258.34	Cholinesterase inhib
Demeton-S-methyl [919-86-8] (1998)	0.05 mg/m³ (IFV)	—	Skin; SEN; A4; BEIA	230.3	Cholinesterase inhib
Diacetone alcohol [123-42-2] (1979)	50 ppm	—	—	116.16	URT & eye irr
Diazinon [333-41-5] (2000)	0.01 mg/m³	—	Skin; A4; BEIA	304.36	Cholinesterase inhib
Diazomethane [334-88-3] (1970)	0.2 ppm	—	A2	42.04	URT & eye irr

Substance [CAS No.] (Documentation date)	TWA	STEL	Notations	MW	TLV® Basis
		ADOPTED VALUES			
Diborane [19287-45-7] (1990)	0.1 ppm	—	—	27.69	URT irr; headache
2-N-Dibutylaminoethanol [102-81-8] (1980)	0.5 ppm	—	Skin; BEI_A	173.29	Eye & URT irr
Dibutyl phenyl phosphate [2528-36-1] (1987)	0.3 ppm	—	Skin; BEI_A	286.26	Cholinesterase inhib; URT irr
Dibutyl phosphate [107-66-4] (2008)	5 mg/m³ (IFV)	—	Skin	210.21	Bladder, eye & URT irr
Dibutyl phthalate [84-74-2] (1990)	5 mg/m³	—	—	278.34	Testicular dam; eye & URT irr
Dichloroacetic acid [79-43-6] (2002)	0.5 ppm	—	Skin; A3	128.95	URT & eye irr; testicular dam
Dichloroacetylene [7572-29-4] (1992)	—	C 0.1 ppm	A3	94.93	Nausea; PNS impair
o-Dichlorobenzene [95-50-1] (1990)	25 ppm	50 ppm	A4	147.01	URT & eye irr; liver dam
p-Dichlorobenzene [106-46-7] (1990)	10 ppm	—	A3	147.01	Eye irr; kidney dam
3,3'-Dichlorobenzidine [91-94-1] (1990)	—(L)	—	Skin; A3	253.13	Bladder cancer; eye irr
1,4-Dichloro-2-butene [764-41-0] (1990)	0.005 ppm	—	Skin; A2	124.99	URT & eye irr
Dichlorodifluoromethane [75-71-8] (1979)	1000 ppm	—	A4	120.91	Card sens
1,3-Dichloro-5,5-dimethyl hydantoin [118-52-5] (1979)	0.2 mg/m³	0.4 mg/m³	—	197.03	URT irr
1,1-Dichloroethane [75-34-3] (1990)	100 ppm	—	A4	98.97	URT & eye irr; liver & kidney dam
1,2-Dichloroethylene, all isomers [540-59-0; 156-59-2; 156-60-5] (1990)	200 ppm	—	—	96.95	CNS impair; eye irr

Substance [CAS No.] (*Documentation* date)	ADOPTED VALUES			Notations	MW	TLV® Basis
	TWA	STEL				
Dichloroethyl ether [111-44-4] (1985)	5 ppm	10 ppm		Skin; A4	143.02	URT & eye irr; nausea
Dichlorofluoromethane [75-43-4] (1977)	10 ppm	—		—	102.92	Liver dam
Dichloromethane [75-09-2] (1997)	50 ppm	—		A3; BEI	84.93	COHb-emia; CNS impair
1,1-Dichloro-1-nitroethane [594-72-9] (1978)	2 ppm	—		—	143.96	URT irr
1,3-Dichloropropene [542-75-6] (2003)	1 ppm	—		Skin; A3	110.98	Kidney dam
2,2-Dichloropropionic acid [75-99-0] (1997)	5 mg/m³ (I)	—		A4	142.97	Eye & URT irr
Dichlorotetrafluoroethane [76-14-2] (1979)	1000 ppm	—		A4	170.93	Pulm func
Dichlorvos (DDVP) [62-73-7] (1998)	0.1 mg/m³ (IFV)	—		Skin; SEN; A4; BEI$_A$	220.98	Cholinesterase inhib
Dicrotophos [141-66-2] (1998)	0.05 mg/m³ (IFV)	—		Skin; A4; BEI$_A$	237.21	Cholinesterase inhib
Dicyclopentadiene [77-73-6] (1973)	5 ppm	—		—	132.21	URT, LRT, & eye irr
Dicyclopentadienyl iron, as Fe [102-54-5] (1990)	10 mg/m³	—		—	186.03	Liver dam
* Dieldrin [60-57-1] (2009)	0.1 mg/m³ (IFV)	—		Skin; A3	380.93	Liver dam; repro eff; CNS impair
Diesel fuel [68334-30-5; 68476-30-2; 68476-31-3; 68476-34-6; 77650-28-3] (2007) as total hydrocarbons	100 mg/m³ (IFV)	—		Skin; A3	Varies	Dermatitis

TLV®–CS

Substance [CAS No.] (Documentation date)	ADOPTED VALUES					
	TWA	STEL	Notations	MW	TLV® Basis	
Diethanolamine [111-42-2] (2008)	1 mg/m³ (IFV)	—	Skin; A3	105.14	Liver & kidney dam	
Diethylamine [109-89-7] (1992)	5 ppm	15 ppm	Skin; A4	73.14	URT & eye irr	
2-Diethylaminoethanol [100-37-8] (1991)	2 ppm	—	Skin	117.19	URT irr; CNS convul	
Diethylene triamine [111-40-0] (1985)	1 ppm	—	Skin	103.17	URT & eye irr	
Di(2-ethylhexyl)phthalate (DEHP) [117-81-7] (1996)	5 mg/m³	—	A3	390.54	LRT irr	
Diethyl ketone [96-22-0] (1995)	200 ppm	300 ppm	—	86.13	URT irr; CNS impair	
Diethyl phthalate [84-66-2] (1996)	5 mg/m³	—	A4	222.23	URT irr	
Difluorodibromomethane [75-61-6] (1962)	100 ppm	—	—	209.83	URT irr; CNS impair; liver dam	
Diglycidyl ether (DGE) [2238-07-5] (2006)	0.01 ppm	—	A4	130.14	Eye & skin irr; male repro dam	
Diisobutyl ketone [108-83-8] (1979)	25 ppm	—	—	142.23	URT & eye irr	
Diisopropylamine [108-18-9] (1979)	5 ppm	—	Skin	101.19	URT irr; eye dam	
N,N-Dimethylacetamide [127-19-5] (1990)	10 ppm	—	Skin; A4; BEI	87.12	Liver dam; embryo/fetal dam	
Dimethylamine [124-40-3] (1989)	5 ppm	15 ppm	A4	45.08	URT irr; GI dam	
Dimethylaniline [121-69-7] (1990)	5 ppm	10 ppm	Skin; A4; BEI_M	121.18	MeHb-emia	
Dimethyl carbamoyl chloride [79-44-7] (2006)	0.005 ppm	—	Skin; A2	107.54	Nasal cancer; URT irr	

Substance [CAS No.] (*Documentation date*)	ADOPTED VALUES			MW	TLV® Basis
	TWA	STEL	Notations		
Dimethyl disulfide [624-92-0] (2006)	0.5 ppm	—	Skin	94.2	URT irr; CNS impair
Dimethylethoxysilane [14857-34-2] (1991)	0.5 ppm	1.5 ppm	—	104.20	URT & eye irr; headache
Dimethylformamide [68-12-2] (1979)	10 ppm	—	Skin; A4; BEI	73.09	Liver dam
1,1-Dimethylhydrazine [57-14-7] (1993)	0.01 ppm	—	Skin; A3	60.12	URT irr; nasal cancer
Dimethyl phthalate [131-11-3] (2005)	5 mg/m³	—	—	194.19	Eye & URT irr
Dimethyl sulfate [77-78-1] (1985)	0.1 ppm	—	Skin; A3	126.10	Eye & skin irr
Dimethyl sulfide [75-18-3] (2001)	10 ppm	—	—	62.14	URT irr
Dinitrobenzene, all isomers [528-29-0; 99-65-0; 100-25-4; 25154-54-5] (1979)	0.15 ppm	—	Skin; BEI_M	168.11	MeHb-emia; eye dam
Dinitro-o-cresol [534-52-1] (1979)	0.2 mg/m³	—	Skin	198.13	Basal metab
3,5-Dinitro-o-toluamide [148-01-6] (2006)	1 mg/m³	—	A4	225.16	Liver dam
Dinitrotoluene [25321-14-6] (1993)	0.2 mg/m³	—	Skin; A3; BEI_M	182.15	Card impair; repro eff
1,4-Dioxane [123-91-1] (1996)	20 ppm	—	Skin; A3	88.10	Liver dam
Dioxathion [78-34-2] (2001)	0.1 mg/m³ (IFV)	—	Skin; A4; BEI_A	456.54	Cholinesterase inhib
1,3-Dioxolane [646-06-0] (1997)	20 ppm	—	—	74.08	Hematologic eff

ADOPTED VALUES

Substance [CAS No.] (*Documentation date*)	TWA	STEL	Notations	MW	TLV® Basis
Diphenylamine [122-39-4] (1990)	10 mg/m³	—	A4	169.24	Liver & kidney dam; hematologic eff
Dipropyl ketone [123-19-3] (1978)	50 ppm	—	—	114.80	URT irr
Diquat [2764-72-9; 85-00-7; 6385-62-2] (1990)	0.5 mg/m³ (I) 0.1 mg/m³ (R)	— —	Skin; A4 Skin; A4	Varies	LRT irr; cataract LRT irr; cataract
Disulfiram [97-77-8] (1979)	2 mg/m³	—	A4	296.54	Vasodilation; nausea
Disulfoton [298-04-4] (2000)	0.05 mg/m³ (IFV)	—	Skin; A4; BEI_A	274.38	Cholinesterase inhib
Diuron [330-54-1] (1974)	10 mg/m³	—	A4	233.10	URT irr
Divinyl benzene [1321-74-0] (1990)	10 ppm	—	—	130.19	URT irr
Dodecyl mercaptan [112-55-0] (2001)	0.1 ppm	—	SEN	202.4	URT irr
Endosulfan [115-29-7] (2008)	0.1 mg/m³ (IFV)	—	Skin; A4	406.95	LRT irr; liver & kidney dam
Endrin [72-20-8] (1979)	0.1 mg/m³	—	Skin; A4	380.93	Liver dam; CNS impair; headache
Enflurane [13838-16-9] (1979)	75 ppm	—	A4	184.50	CNS impair; card impair
Epichlorohydrin [106-89-8] (1994)	0.5 ppm	—	Skin; A3	92.53	URT irr; male repro
EPN [2104-64-5] (2000)	0.1 mg/m³ (I)	—	Skin; A4; BEI_A	323.31	Cholinesterase inhib
Ethane [74-84-0]	See Aliphatic hydrocarbon gases: Alkanes [C_1–C_4]				

Substance [CAS No.] (*Documentation* date)	ADOPTED VALUES			MW	TLV® Basis
	TWA	STEL	Notations		
Ethanol [64-17-5] (2008)	—	1000 ppm	A3	46.07	URT irr
Ethanolamine [141-43-5] (1985)	3 ppm	6 ppm	—	61.08	Eye & skin irr
Ethion [563-12-2] (2000)	0.05 mg/m³ (IFV)	—	Skin; A4; BEI$_A$	384.48	Cholinesterase inhib
2-Ethoxyethanol (EGEE) [110-80-5] (1981)	5 ppm	—	Skin; BEI	90.12	Male repro dam; embryo/fetal dam
2-Ethoxyethyl acetate (EGEEA) [111-15-9] (1981)	5 ppm	—	Skin; BEI	132.16	Male repro dam
Ethyl acetate [141-78-6] (1979)	400 ppm	—	—	88.10	URT & eye irr
Ethyl acrylate [140-88-5] (1986)	5 ppm	15 ppm	A4	100.11	URT, eye, & GI irr; CNS impair; skin sens
Ethylamine [75-04-7] (1991)	5 ppm	15 ppm	Skin	45.08	Eye & skin irr, eye dam
Ethyl amyl ketone [541-85-5] (2006)	10 ppm	—	—	128.21	Neurotoxicity
‡ Ethyl benzene [100-41-4] (1998)	(100 ppm)	(125 ppm)	A3; BEI	106.16	(URT & eye irr; CNS impair)
Ethyl bromide [74-96-4] (1990)	5 ppm	—	Skin; A3	108.98	Liver dam; CNS impair
Ethyl tert-butyl ether (ETBE) [637-92-3] (1997)	5 ppm	—	—	102.18	Pulm func; testicular dam
Ethyl butyl ketone [106-35-4] (1995)	50 ppm	75 ppm	—	114.19	CNS impair; eye & skin irr
Ethyl chloride [75-00-3] (1992)	100 ppm	—	Skin; A3	64.52	Liver dam

TLV®–CS

Substance [CAS No.] (*Documentation* date)	ADOPTED VALUES TWA	ADOPTED VALUES STEL	Notations	MW	TLV® Basis
Ethyl cyanoacrylate [7085-85-0] (1995)	0.2 ppm	—	—	125.12	URT & skin irr
Ethylene [74-85-1] (2001)	200 ppm	—	A4	28.05	Asphyxia
Ethylene chlorohydrin [107-07-3] (1985)	—	C 1 ppm	Skin; A4	80.52	CNS impair; liver & kidney dam
Ethylenediamine [107-15-3] (1990)	10 ppm	—	Skin; A4	60.10	
Ethylene dibromide [106-93-4] (1980)	—	—	Skin; A3	187.88	
Ethylene dichloride [107-06-2] (1977)	10 ppm	—	A4	98.96	Liver dam; nausea
Ethylene glycol [107-21-1] (1992)	—	C 100 mg/m³ (H)	A4	62.07	URT & eye irr
Ethylene glycol dinitrate (EGDN) [628-96-6] (1980)	0.05 ppm	—	Skin	152.06	Vasodilation; headache
Ethylene oxide [75-21-8] (1990)	1 ppm	—	A2	44.05	Cancer; CNS impair
Ethyleneimine [151-56-4] (2008)	0.05 ppm	0.1 ppm	Skin; A3	43.08	URT irr, liver & kidney dam
Ethyl ether [60-29-7] (1966)	400 ppm	500 ppm	—	74.12	CNS impair; URT irr
Ethyl formate [109-94-4] (1970)	100 ppm	—	—	74.08	URT & eye irr
2-Ethylhexanoic acid [149-57-5] (2006)	5 mg/m³ (IFV)	—	—	144.24	Teratogenic eff
Ethylidene norbornene [16219-75-3] (1971)	—	C 5 ppm	—	120.19	URT & eye irr
Ethyl mercaptan [75-08-1] (2003)	0.5 ppm	—	—	62.13	URT irr, CNS impair

	ADOPTED VALUES				
Substance [CAS No.] (*Documentation date*)	TWA	STEL	Notations	MW	TLV® Basis
N-Ethylmorpholine [100-74-3] (1985)	5 ppm	—	Skin	115.18	URT irr; eye dam
Ethyl silicate [78-10-4] (1979)	10 ppm	—	—	208.30	URT & eye irr; kidney dam
Fenamiphos [22224-92-6] (2005)	0.05 mg/m³ (IFV)	—	Skin; A4; BEI$_A$	303.40	Cholinesterase inhib
Fensulfothion [115-90-2] (2004)	0.01 mg/m³ (IFV)	—	Skin; A4; BEI$_A$	308.35	Cholinesterase inhib
Fenthion [55-38-9] (2005)	0.05 mg/m³ (IFV)	—	Skin; A4; BEI$_A$	278.34	Cholinesterase inhib
Ferbam [14484-64-1] (2008)	5 mg/m³ (I)	—	A4	416.50	CNS impair; body weight eff; spleen dam
Ferrovanadium dust [12604-58-9] (1990)	1 mg/m³	3 mg/m³	—	—	Eye, URT, & LRT irr
Flour dust (2001)	0.5 mg/m³ (I)	—	SEN	—	Asthma; URT irr, bronchitis
Fluorides, as F (1979)	2.5 mg/m³	—	A4; BEI	Varies	Bone dam; fluorosis
Fluorine [7782-41-4] (1970)	1 ppm	2 ppm	—	38.00	URT, eye, & skin irr
Fonofos [944-22-9] (2005)	0.01 mg/m³ (IFV)	—	Skin; A4; BEI$_A$	246.32	Cholinesterase inhib
Formaldehyde [50-00-0] (1987)	—	C 0.3 ppm	SEN; A2	30.03	URT & eye irr
Formamide [75-12-7] (1985)	10 ppm	—	Skin	45.04	Eye & skin irr; kidney & liver dam
Formic acid [64-18-6] (1965)	5 ppm	10 ppm	—	46.02	URT, eye, & skin irr

Substance [CAS No.] *(Documentation date)*	ADOPTED VALUES					TLV® Basis
	TWA	STEL	Notations	MW		
Furfural [98-01-1] (1978)	2 ppm	—	Skin; A3; BEI	96.08		URT & eye irr
Furfuryl alcohol [98-00-0] (1979)	10 ppm	15 ppm	Skin	98.10		URT & eye irr
Gallium arsenide [1303-00-0] (2004)	0.0003 mg/m³ (R)	—	A3	144.64		LRT irr
Gasoline [86290-81-5] (1990)	300 ppm	500 ppm	A3	—		URT & eye irr; CNS impair
Germanium tetrahydride [7782-65-2] (1970)	0.2 ppm	—	—	76.63		Hematologic eff
Glutaraldehyde [111-30-8], activated and inactivated (1998)	—	C 0.05 ppm	SEN; A4	100.11		URT, skin, & eye irr; CNS impair
Glycerin mist [56-81-5] (1990)	10 mg/m³	—	—	92.09		URT irr
Glycidol [556-52-5] (1993)	2 ppm	—	A3	74.08		URT, eye, & skin irr
Glyoxal [107-22-2] (1999)	0.1 mg/m³ (IFV)	—	SEN; A4	58.04		URT irr; larynx metaplasia
Grain dust (oat, wheat, barley) (1985)	4 mg/m³	—	—	NA		Bronchitis; URT irr; pulm func
Graphite (all forms except graphite fibers) [7782-42-5] (1988)	2 mg/m³ (R)	—	—	—		Pneumoconiosis
Hafnium [7440-58-6] and compounds, as Hf (1990)	0.5 mg/m³	—	—	178.49		URT & eye irr; liver dam
Halothane [151-67-7] (1979)	50 ppm	—	A4	197.39		Liver dam; CNS impair; vasodilation
Helium [7440-59-7] (1990)	Simple asphyxiant (D)			4.00		Asphyxia

Substance [CAS No.] (*Documentation date*)	ADOPTED VALUES				TLV® Basis
	TWA	STEL	Notations	MW	
Heptachlor [76-44-8] and Heptachlor epoxide [1024-57-3] (1990)	0.05 mg/m³	—	Skin; A3	373.32 389.40	Liver dam
Heptane, all isomers [142-82-5; 590-35-2; 565-59-3; 108-08-7; 591-76-4; 589-34-4] (1979)	400 ppm	500 ppm	—	100.20	CNS impair; URT irr
Hexachlorobenzene [118-74-1] (1994)	0.002 mg/m³	—	Skin; A3	284.78	Porphyrin eff; skin dam; CNS impair
Hexachlorobutadiene [87-68-3] (1979)	0.02 ppm	—	Skin; A3	260.76	Kidney dam
Hexachlorocyclopentadiene [77-47-4] (1990)	0.01 ppm	—	A4	272.75	URT irr
Hexachloroethane [67-72-1] (1990)	1 ppm	—	Skin; A3	236.74	Liver & kidney dam
Hexachloronaphthalene [1335-87-1] (1965)	0.2 mg/m³	—	Skin	334.74	Liver dam; chloracne
Hexafluoroacetone [684-16-2] (1986)	0.1 ppm	—	Skin	166.02	Testicular dam; kidney dam
Hexafluoropropylene [116-15-4] (2009)	0.1 ppm	—	—	150.02	Kidney dam
Hexahydrophthalic anhydride, all isomers [85-42-7; 13149-00-3; 14166-21-3] (2002)	—	C 0.005 mg/m³ (IFV)	SEN	154.17	Resp sens; eye, skin, & URT irr
Hexamethylene diisocyanate [822-06-0] (1985)	0.005 ppm	—	—	168.22	URT irr; resp sens
Hexamethyl phosphoramide [680-31-9] (1990)	—	—	Skin; A3	179.20	URT cancer

TLV®-CS

ADOPTED VALUES

Substance [CAS No.] (*Documentation date*)	TWA	STEL	Notations	MW	TLV® Basis
n-Hexane [110-54-3] (1996)	50 ppm	—	Skin; BEI	86.18	CNS impair; peripheral neuropathy; eye irr
Hexane, other isomers (1979)	500 ppm	1000 ppm	—	86.18	CNS impair; URT & eye irr
1,6-Hexanediamine [124-09-4] (1990)	0.5 ppm	—	—	116.21	URT & skin irr
1-Hexene [592-41-6] (1999)	50 ppm	—	—	84.16	CNS impair
sec-Hexyl acetate [108-84-9] (1963)	50 ppm	—	—	144.21	Eye & URT irr
Hexylene glycol [107-41-5] (1974)	—	C 25 ppm	—	118.17	Eye & URT irr
Hydrazine [302-01-2] (1988)	0.01 ppm	—	Skin; A3	32.05	URT cancer
Hydrogen [1333-74-0] (1990)	Simple asphyxiant (D)			1.01	Asphyxia
Hydrogenated terphenyls (nonirradiated) [61788-32-7] (1990)	0.5 ppm	—	—	241.00	Liver dam
Hydrogen bromide [10035-10-6] (2001)	—	C 2 ppm	—	80.92	URT irr
Hydrogen chloride [7647-01-0] (2000)	—	C 2 ppm	A4	36.47	URT irr
Hydrogen cyanide and cyanide salts, as CN (1991)					URT irr; headache; nausea; thyroid eff
Hydrogen cyanide [74-90-8]	—	C 4.7 ppm	Skin	27.03	
Cyanide salts [592-01-8; 151-50-8; 143-33-9]	—	C 5 mg/m³	Skin	Varies	

Substance [CAS No.] (Documentation date)	ADOPTED VALUES			Notations	MW	TLV® Basis
	TWA	STEL				
Hydrogen fluoride [7664-39-3], as F (2004)	0.5 ppm	C 2 ppm		Skin; BEI	20.01	URT, LRT, skin & eye irr; fluorosis
Hydrogen peroxide [7722-84-1] (1990)	1 ppm	—		A3	34.02	Eye, URT & skin irr
Hydrogen selenide [7783-07-5], as Se (1990)	0.05 ppm	—		—	80.98	URT & eye irr; nausea
* Hydrogen sulfide [7783-06-4] (2009)	1 ppm	5 ppm		—	34.08	URT irr; CNS impair
Hydroquinone [123-31-9] (2007)	1 mg/m³	—		SEN; A3	110.11	Eye irr; eye dam
2-Hydroxypropyl acrylate [999-61-1] (1997)	0.5 ppm	—		Skin; SEN	130.14	Eye & URT irr
Indene [95-13-6] (2007)	5 ppm	—		—	116.15	Liver dam
Indium [7440-74-6] and compounds, as In (1990)	0.1 mg/m³	—		—	49.00	Pulm edema; pneumonitis; dental erosion; malaise
Iodine and iodides (2007)						
Iodine [7553-56-2]	0.01 ppm (IFV)	0.1 ppm (V)		A4	Varies	Hypothyroidism; URT irr
Iodides	0.01 ppm (IFV)	—		A4	Varies	Hypothyroidism; URT irr
Iodoform [75-47-8] (1979)	0.6 ppm	—		—	393.78	CNS impair
Iron oxide (Fe₂O₃) [1309-37-1] (2005)	5 mg/m³ (R)	—		A4	159.70	Pneumoconiosis
Iron pentacarbonyl [13463-40-6], as Fe (1979)	0.1 ppm	0.2 ppm		—	195.90	Pulm edema; CNS impair

Substance [CAS No.] (*Documentation date*)	ADOPTED VALUES				MW	TLV® Basis
	TWA	STEL	Notations			
Iron salts, soluble, as Fe (1990)	1 mg/m³	—	—		Varies	URT & skin irr
Isoamyl alcohol [123-51-3] (1990)	100 ppm	125 ppm	—		88.15	Eye & URT irr
Isobutanol [78-83-1] (1973)	50 ppm	—	—		74.12	Skin & eye irr
Isobutyl acetate [110-19-0] (1966)	150 ppm	—	—		116.16	Eye & URT irr
Isobutyl nitrite [542-56-3] (2000)	—	C 1 ppm (IFV)	A3; BE$_M$		103.12	Vasodilation; MeHb-emia
Isooctyl alcohol [26952-21-6] (1990)	50 ppm	—	Skin		130.23	URT irr
Isophorone [78-59-1] (1992)	—	C 5 ppm	A3		138.21	Eye & URT irr, CNS impair; malaise; fatigue
Isophorone diisocyanate [4098-71-9] (1985)	0.005 ppm	—	—		222.30	Resp sens
Isopropanol [67-63-0]	Name change; see 2-Propanol					
2-Isopropoxyethanol [109-59-1] (1990)	25 ppm	—	Skin		104.15	Hematologic eff
Isopropyl acetate [108-21-4] (2001)	100 ppm	200 ppm	—		102.13	Eye & URT irr; CNS impair
Isopropylamine [75-31-0] (1962)	5 ppm	10 ppm	—		59.08	URT irr; eye dam
N-Isopropylaniline [768-52-5] (1990)	2 ppm	—	Skin; BE$_M$		135.21	MeHb-emia
Isopropyl ether [108-20-3] (1979)	250 ppm	310 ppm	—		102.17	Eye & URT irr

| Substance [CAS No.] (Documentation date) | ADOPTED VALUES | | | | TLV® Basis |
	TWA	STEL	Notations	MW	
Isopropyl glycidyl ether (IGE) [4016-14-2] (1979)	50 ppm	75 ppm	—	116.18	URT & eye irr; dermatitis
Kaolin [1332-58-7] (1990)	2 mg/m³ (E,R)	—	A4	—	Pneumoconiosis
Kerosene [8008-20-6; 64742-81-0]/Jet fuels, as total hydrocarbon vapor (2003)	200 mg/m³ (P)	—	Skin; A3	Varies	Skin & URT irr; CNS impair
Ketene [463-51-4] (1962)	0.5 ppm	1.5 ppm	—	42.04	URT irr; pulm edema
Lead [7439-92-1] and inorganic compounds, as Pb (1991)	0.05 mg/m³	—	A3; BEI	207.20	CNS & PNS impair; hematologic eff
Lead chromate [7758-97-6], as Pb (1990)	0.05 mg/m³	—	A2; BEI	323.22	Male repro dam; teratogenic eff; vasoconstriction
as Cr	0.012 mg/m³	—	A2	Varies	
Lindane [58-89-9] (1990)	0.5 mg/m³	—	Skin; A3	290.85	Liver dam; CNS impair
Lithium hydride [7580-67-8] (1990)	0.025 mg/m³	—	—	7.95	Skin, eye, & URT irr
L.P.G. (Liquefied petroleum gas) [68476-85-7]	See Aliphatic hydrocarbon gases: Alkanes [C₁–C₄]				
Magnesium oxide [1309-48-4] (2000)	10 mg/m³ (I)	—	A4	40.32	
Malathion [121-75-5] (2000)	1 mg/m³ (IFV)	—	Skin; A4; BEI_A	330.36	Cholinesterase inhib
‡ Maleic anhydride [108-31-6] (1997)	(0.1 ppm)	—	SEN; A4	98.06	(Eye, URT, & skin irr)

ADOPTED VALUES

Substance [CAS No.] (*Documentation date*)	TWA	STEL	Notations	MW	TLV® Basis
‡ (Manganese [7439-96-5] and inorganic compounds, as Mn) (1992)	(0.2 mg/m³)	—	(—)	54.94 Varies	(CNS impair)
Manganese cyclopentadienyl tricarbonyl [12079-65-1], as Mn (1992)	0.1 mg/m³	—	Skin	204.10	Skin irr; CNS impair
Mercury [7439-97-6], as Hg (1991)				200.59	
Alkyl compounds (1992)	0.01 mg/m³	0.03 mg/m³	Skin	Varies	CNS & PNS impair; kidney dam
Aryl compounds	0.1 mg/m³	—	Skin	Varies	CNS impair; kidney dam
Elemental and inorganic forms	0.025 mg/m³	—	Skin; A4; BEI	Varies	CNS impair; kidney dam
Mesityl oxide [141-79-7] (1992)	15 ppm	25 ppm	—	98.14	Eye & URT irr; CNS impair
Methacrylic acid [79-41-4] (1992)	20 ppm	—	—	86.09	Skin & eye irr
Methane [74-82-8]	See Aliphatic hydrocarbon gases: Alkanes [C₁–C₄]				
Methanol [67-56-1] (2008)	200 ppm	250 ppm	Skin; BEI	32.04	Headache; eye dam
Methomyl [16752-77-5] (1992)	2.5 mg/m³	—	A4; BEI_A	162.20	Cholinesterase inhib
Methoxychlor [72-43-5] (1992)	10 mg/m³	—	A4	345.65	Liver dam; CNS impair
2-Methoxyethanol (EGME) [109-86-4] (2005)	0.1 ppm	—	Skin	76.09	Hematologic eff; repro eff
2-Methoxyethyl acetate (EGMEA) [110-49-6] (2005)	0.1 ppm	—	Skin	118.13	Hematologic eff; repro eff

| Substance [CAS No.] (*Documentation date*) | ADOPTED VALUES | | | MW | TLV® Basis |
	TWA	STEL	Notations		
(2-Methoxymethylethoxy)propanol (DPGME) [34590-94-8] (1979)	100 ppm	150 ppm	Skin	148.20	Eye & URT irr; CNS impair
4-Methoxyphenol [150-76-5] (1992)	5 mg/m³	—	—	124.15	Eye irr; skin dam
1-Methoxy-2-propanol (PGME) [107-98-2] (1992)	100 ppm	150 ppm	—	90.12	Eye irr; CNS impair
Methyl acetate [79-20-9] (1992)	200 ppm	250 ppm	—	74.08	Headache; eye & URT irr; ocular nerve dam
Methyl acetylene [74-99-7] (1956)	1000 ppm	—	—	40.07	CNS impair
Methyl acetylene-propadiene mixture (MAPP) [59355-75-8] (1964)	1000 ppm	1250 ppm	—	40.07	CNS impair
Methyl acrylate [96-33-3] (1997)	2 ppm	—	Skin; SEN; A4	86.09	Eye, skin, & URT irr; eye dam
‡ Methylacrylonitrile [126-98-7] (1992)	1 ppm	—	Skin; ( )	67.09	CNS impair; eye & skin irr
Methylal [109-87-5] (1970)	1000 ppm	—	—	76.10	Eye irr; CNS impair
Methylamine [74-89-5] (1990)	5 ppm	15 ppm	—	31.06	Eye, skin, & URT irr
Methyl n-amyl ketone [110-43-0] (1978)	50 ppm	—	—	114.18	Eye & skin irr
N-Methyl aniline [100-61-8] (1992)	0.5 ppm	—	Skin; BEI_M	107.15	MeHb-emia; CNS impair
Methyl bromide [74-83-9] (1994)	1 ppm	—	Skin; A4	94.95	URT & skin irr

| Substance [CAS No.] (*Documentation date*) | ADOPTED VALUES | | | MW | TLV® Basis |
	TWA	STEL	Notations		
Methyl tert-butyl ether (MTBE) [1634-04-4] (1999)	50 ppm	—	A3	88.17	URT irr; kidney dam
Methyl n-butyl ketone [591-78-6] (1995)	5 ppm	10 ppm	Skin; BEI	100.16	Peripheral neuropathy; testicular dam
Methyl chloride [74-87-3] (1992)	50 ppm	100 ppm	Skin; A4	50.49	CNS impair; liver & kidney dam; testicular dam; teratogenic eff
Methyl chloroform [71-55-6] (1992)	350 ppm	450 ppm	A4; BEI	133.42	CNS impair; liver dam
Methyl 2-cyanoacrylate [137-05-3] (1995)	0.2 ppm	—	—	111.10	URT & eye irr
Methylcyclohexane [108-87-2] (1962)	400 ppm	—	—	98.19	URT irr, CNS impair; liver & kidney dam
Methylcyclohexanol [25639-42-3] (2005)	50 ppm	—	—	114.19	URT & eye irr
o-Methylcyclohexanone [583-60-8] (1970)	50 ppm	75 ppm	Skin	112.17	URT & eye irr; CNS impair
2-Methylcyclopentadienyl manganese tricarbonyl [12108-13-3], as Mn (1970)	0.2 mg/m³	—	Skin	218.10	CNS impair; lung, liver, & kidney dam
Methyl demeton [8022-00-2] (2006)	0.05 mg/m³ (IFV)	—	Skin; BEI_A	230.30	Cholinesterase inhib
Methylene bisphenyl isocyanate (MDI) [101-68-8] (1985)	0.005 ppm	—	—	250.26	Resp sens
4,4'-Methylene bis(2-chloroaniline) [MBOCA; MOCA®] [101-14-4] (1991)	0.01 ppm	—	Skin; A2; BEI	267.17	Bladder cancer; MeHb-emia

TLV®–CS

Substance [CAS No.] (Documentation date)	ADOPTED VALUES			MW	TLV® Basis
	TWA	STEL	Notations		
Methylene bis(4-cyclohexylisocyanate) [5124-30-1] (1985)	0.005 ppm	—	—	262.35	Resp sens; LRT irr
4,4'-Methylene dianiline [101-77-9] (1992)	0.1 ppm	—	Skin; A3	198.26	Liver dam
Methyl ethyl ketone (MEK) [78-93-3] (1992)	200 ppm	300 ppm	BEI	72.10	URT irr; CNS & PNS impair
Methyl ethyl ketone peroxide [1338-23-4] (1992)	—	C 0.2 ppm	—	176.24	Eye & skin irr; liver & kidney dam
Methyl formate [107-31-3] (1962)	100 ppm	150 ppm	—	60.05	URT, LRT, & eye irr
Methyl hydrazine [60-34-4] (1991)	0.01 ppm	—	Skin; A3	46.07	URT & eye irr; lung cancer; liver dam
Methyl iodide [74-88-4] (1978)	2 ppm	—	Skin	141.95	Eye dam; CNS impair
Methyl isoamyl ketone [110-12-3] (1979)	50 ppm	—	—	114.20	URT & eye irr; kidney & liver dam; CNS impair
Methyl isobutyl carbinol [108-11-2] (1966)	25 ppm	40 ppm	Skin	102.18	URT & eye irr; CNS impair
* Methyl isobutyl ketone [108-10-1] (2009)	20 ppm	75 ppm	A3; BEI	100.16	URT irr, dizziness; headache
Methyl isocyanate [624-83-9] (1986)	0.02 ppm	—	Skin	57.05	URT irr
‡ Methyl isopropyl ketone [563-80-4] (1992)	(200 ppm)	—	—	86.14	(URT & eye irr)
Methyl mercaptan [74-93-1] (2003)	0.5 ppm	—	—	48.11	Liver dam
Methyl methacrylate [80-62-6] (1992)	50 ppm	100 ppm	SEN; A4	100.13	URT & eye irr; body weight eff; pulm edema

TLV®–CS

ADOPTED VALUES

Substance [CAS No.] (Documentation date)	TWA	STEL	Notations	MW	TLV® Basis
1-Methyl naphthalene [90-12-0] and 2-Methyl naphthalene [91-57-6] (2006)	0.5 ppm	—	Skin; A4	142.2	LRT irr; lung dam
Methyl parathion [298-00-0] (2008)	0.02 mg/m³ (IFV)	—	Skin; A4; BEI_A	263.2	Cholinesterase inhib
Methyl propyl ketone [107-87-9] (2006)	—	150 ppm	—	86.17	Pulm func; eye irr
Methyl silicate [681-84-5] (1978)	1 ppm	—	—	152.22	URT irr; eye dam
* α-Methyl styrene [98-83-9] (2009)	10 ppm	—	A3	118.18	URT irr; kidney dam; female repro dam
Methyl vinyl ketone [78-94-4] (1994)	—	C 0.2 ppm	Skin; SEN	70.10	URT & eye irr; CNS impair
Metribuzin [21087-64-9] (1981)	5 mg/m³	—	A4	214.28	Liver dam; hematologic eff
Mevinphos [7786-34-7] (1998)	0.01 mg/m³ (IFV)	—	Skin; A4; BEI_A	224.16	Cholinesterase inhib
Mica [12001-26-2] (1962)	3 mg/m³ (R)	—	—	—	Pneumoconiosis
* Mineral oil [8012-95-1], excluding metal working fluids				—	URT irr
Pure, highly and severely refined	5 mg/m³ (I)	—	A4		
Poorly and mildly refined	— (L)	—	A2		

Substance [CAS No.] (*Documentation date*)	ADOPTED VALUES				
	TWA	STEL	Notations	MW	TLV® Basis
Molybdenum [7439-98-7], as Mo (1999)				95.95	
Soluble compounds	0.5 mg/m³ (R)	—	A3		LRT irr
Metal and insoluble compounds	10 mg/m³ (I)	—	—		
	3 mg/m³ (R)	—	—		
Monochloroacetic acid [79-11-8] (2005)	0.5 ppm (IFV)	—	Skin; A4	94.5	URT irr
Monocrotophos [6923-22-4] (2002)	0.05 mg/m³ (IFV)	—	Skin; A4; BEI$_A$	223.16	Cholinesterase inhib
Morpholine [110-91-8] (1992)	20 ppm	—	Skin; A4	87.12	Eye dam; URT irr
Naled [300-76-5] (2002)	0.1 mg/m³ (IFV)	—	Skin; SEN; A4; BEI$_A$	380.79	Cholinesterase inhib
Naphthalene [91-20-3] (1992)	10 ppm	15 ppm	Skin; A4	128.19	Hematologic eff; URT & eye irr; eye dam
β-Naphthylamine [91-59-8] (1979)	— (L)	—	A1	143.18	Bladder cancer
Natural gas [8006-14-2]	See Aliphatic hydrocarbon gases: Alkanes [C$_1$–C$_4$]				
Natural rubber latex [9006-04-6], as inhalable allergenic proteins (2007)	0.0001 mg/m³ (I)	—	Skin; SEN	Varies	Resp sens
Neon [7440-01-9] (1992)	Simple asphyxiant (D)			20.18	Asphyxia

Substance [CAS No.] *(Documentation date)*	ADOPTED VALUES				TLV® Basis
	TWA	STEL	Notations	MW	
Nickel, as Ni (1996)					
Elemental [7440-02-0]	1.5 mg/m³ (I)	—	A5	58.71	Dermatitis; pneumoconiosis
Soluble inorganic compounds (NOS)	0.1 mg/m³ (I)	—	A4	Varies	Lung dam; nasal cancer
Insoluble inorganic compounds (NOS)	0.2 mg/m³ (I)	—	A1	Varies	Lung cancer
Nickel subsulfide [12035-72-2], as Ni	0.1 mg/m³ (I)	—	A1	240.19	Lung cancer
Nickel carbonyl [13463-39-3], as Ni (1980)	0.05 ppm	—	—	170.73	Lung & nasal cancer
Nicotine [54-11-5] (1992)	0.5 mg/m³	—	Skin	162.23	GI dam; CNS impair; card impair
Nitrapyrin [1929-82-4] (1992)	10 mg/m³	20 mg/m³	A4	230.93	Liver dam
Nitric acid [7697-37-2] (1992)	2 ppm	4 ppm	—	63.02	URT & eye irr; dental erosion
Nitric oxide [10102-43-9] (1992)	25 ppm	—	BEI$_M$	30.01	Hypoxia/cyanosis; nitrosyl-Hb form; URT irr
p-Nitroaniline [100-01-6] (1992)	3 mg/m³	—	Skin; A4; BEI$_M$	138.12	MeHb-emia; liver dam; eye irr
Nitrobenzene [98-95-3] (1992)	1 ppm	—	Skin; A3; BEI	123.11	MeHb-emia
p-Nitrochlorobenzene [100-00-5] (1985)	0.1 ppm	—	Skin; A3; BEI$_M$	157.56	MeHb-emia
4-Nitrodiphenyl [92-93-3] (1992)	— (L)	—	Skin; A2	199.20	Bladder cancer
Nitroethane [79-24-3] (1979)	100 ppm	—	—	75.07	URT irr; CNS impair; liver dam

	ADOPTED VALUES				
Substance [CAS No.] (*Documentation* date)	TWA	STEL	Notations	MW	TLV® Basis
Nitrogen [7727-37-9] (1992)		Simple asphyxiant [(D)]		14.01	Asphyxia
Nitrogen dioxide [10102-44-0] (1978)	3 ppm	5 ppm	A4	46.01	URT & LRT irr
Nitrogen trifluoride [7783-54-2] (1992)	10 ppm	—	BEI$_M$	71.00	MeHb-emia; liver & kidney dam
Nitroglycerin (NG) [55-63-0] (1980)	0.05 ppm	—	Skin	227.09	Vasodilation
Nitromethane [75-52-5] (1997)	20 ppm	—	A3	61.04	Thyroid eff; URT irr; lung dam
1-Nitropropane [108-03-2] (1992)	25 ppm	—	A4	89.09	URT & eye irr; liver dam
2-Nitropropane [79-46-9] (1992)	10 ppm	—	A3	89.09	Liver dam; liver cancer
N-Nitrosodimethylamine [62-75-9] (1992)	—[(L)]	—	Skin; A3	74.08	Liver & kidney cancer; liver dam
Nitrotoluene, all isomers (1992) [88-72-2; 99-08-1; 99-99-0]	2 ppm	—	Skin; BEI$_M$	137.13	MeHb-emia
5-Nitro-o-toluidine [99-55-8] (2006)	1 mg/m^3 [(I)]	—	A3	152.16	Liver dam
Nitrous oxide [10024-97-2] (1986)	50 ppm	—	A4	44.02	CNS impair; hematologic eff; embryo/fetal dam
Nonane [111-84-2], all isomers (1992)	200 ppm	—	—	128.26	CNS impair
Octachloronaphthalene [2234-13-1] (1970)	0.1 mg/m^3	0.3 mg/m^3	Skin	403.74	Liver dam

| Substance [CAS No.] (*Documentation* date) | ADOPTED VALUES | | | | TLV® Basis |
	TWA	STEL	Notations	MW	
Octane, all isomers [111-65-9] (1979)	300 ppm	—	—	114.22	URT irr
Osmium tetroxide [20816-12-0], as Os (1979)	0.0002 ppm	0.0006 ppm	—	254.20	Eye, URT, & skin irr
Oxalic acid [144-62-7] (1992)	1 mg/m³	2 mg/m³	—	90.04	URT, eye, & skin irr
p,p'-Oxybis(benzenesulfonyl hydrazide) [80-51-3] (1997)	0.1 mg/m³ (I)	—	—	326.00	Teratogenic eff
Oxygen difluoride [7783-41-7] (1983)	—	C 0.05 ppm	—	54.00	Headache; pulm edema; URT irr
Ozone [10028-15-6] (1995)				48.00	Pulm func
Heavy work	0.05 ppm	—	A4		
Moderate work	0.08 ppm	—	A4		
Light work	0.10 ppm	—	A4		
Heavy, moderate, or light workloads (≤ 2 hours)	0.20 ppm	—	A4		
Paraffin wax fume [8002-74-2] (1972)	2 mg/m³	—	—	—	URT irr; nausea
Paraquat [4685-14-7] (1979)	0.5 mg/m³	—	—	257.18	Lung dam
	0.1 mg/m³ (R)	—	—		
Parathion [56-38-2] (2000)	0.05 mg/m³ (IFV)	See Appendix B	Skin; A4; BEI	291.27	Cholinesterase inhib
Particles (insoluble or poorly soluble) not otherwise specified					

| Substance [CAS No.] (Documentation date) | ADOPTED VALUES | | | | TLV® Basis |
	TWA	STEL	Notations	MW	
Pentaborane [19624-22-7] (1970)	0.005 ppm	0.015 ppm	—	63.17	CNS convul; CNS impair
Pentachloronaphthalene [1321-64-8] (1970)	0.5 mg/m³	—	Skin	300.40	Liver dam; chloracne
Pentachloronitrobenzene [82-68-8] (1988)	0.5 mg/m³	—	A4	295.36	Liver dam
Pentachlorophenol [87-86-5] (1992)	0.5 mg/m³	—	Skin; A3; BEI	266.35	URT & eye irr; CNS impair; card impair
Pentaerythritol [115-77-5] (1970)	10 mg/m³	—	—	136.15	Eye & URT irr
Pentane, all isomers [78-78-4; 109-66-0; 463-82-1] (1989)	600 ppm	—	—	72.15	Peripheral neuropathy
Pentyl acetate, all isomers [628-63-7; 626-38-0; 123-92-2; 625-16-1; 624-41-9; 620-11-1] (1997)	50 ppm	100 ppm	—	130.20	URT irr
Perchloromethyl mercaptan [594-42-3] (1988)	0.1 ppm	—	—	185.87	Eye & URT irr
Perchloryl fluoride [7616-94-6] (1962)	3 ppm	6 ppm	—	102.46	LRT & URT irr; MeHb-emia; fluorosis
Perfluorobutyl ethylene [19430-93-4] (2001)	100 ppm	—	—	246.1	Hematologic eff
Perfluoroisobutylene [382-21-8] (1989)	—	C 0.01 ppm	—	200.04	URT irr; hematologic eff
Persulfates, as persulfate (1993)	0.1 mg/m³	—	—	Varies	Skin irr
Phenol [108-95-2] (1992)	5 ppm	—	Skin; A4; BEI	94.11	URT irr; lung dam; CNS impair
Phenothiazine [92-84-2] (1968)	5 mg/m³	—	Skin	199.26	Eye photosens; skin irr

TLV®–CS

Substance [CAS No.] (*Documentation date*)	ADOPTED VALUES				MW	TLV® Basis
	TWA	STEL	Notations			
N-Phenyl-β-naphthylamine [135-88-6] (1992)	—(L)	—	A4		219.29	Cancer
o-Phenylenediamine [95-54-5] (1988)	0.1 mg/m³	—	A3		108.05	Anemia
m-Phenylenediamine [108-45-2] (1988)	0.1 mg/m³	—	A4		108.05	Liver dam; skin irr
p-Phenylenediamine [106-50-3] (1988)	0.1 mg/m³	—	A4		108.05	URT irr, skin sens
Phenyl ether [101-84-8], vapor (1979)	1 ppm	2 ppm	—		170.20	URT & eye irr, nausea
Phenyl glycidyl ether (PGE) [122-60-1] (1992)	0.1 ppm	—	Skin; SEN; A3		150.17	Testicular dam
Phenylhydrazine [100-63-0] (1988)	0.1 ppm	—	Skin; A3		108.14	Anemia; URT & skin irr
Phenyl mercaptan [108-98-5] (2001)	0.1 ppm	—	Skin		110.18	CNS impair; eye & skin irr
Phenylphosphine [638-21-1] (1992)	—	C 0.05 ppm	—		110.10	Dermatitis; hematologic eff; testicular dam
Phorate [298-02-2] (2002)	0.05 mg/m³ (IFV)	—	Skin; A4; BEI_A		260.40	Cholinesterase inhib
Phosgene [75-44-5] (1992)	0.1 ppm	—	—		98.92	URT irr, pulm edema; pulm emphysema
Phosphine [7803-51-2] (1992)	0.3 ppm	1 ppm	—		34.00	URT & GI irr; headache; CNS impair
Phosphoric acid [7664-38-2] (1992)	1 mg/m³	3 mg/m³	—		98.00	URT, eye, & skin irr
Phosphorus (yellow) [12185-10-3] (1992)	0.1 mg/m³	—	—		123.92	LRT, URT, & GI irr, liver dam

Substance [CAS No.] (*Documentation date*)	ADOPTED VALUES			MW	TLV® Basis
	TWA	STEL	Notations		
Phosphorus oxychloride [10025-87-3] (1979)	0.1 ppm	—	—	153.35	URT irr
Phosphorus pentachloride [10026-13-8] (1985)	0.1 ppm	—	—	208.24	URT & eye irr
Phosphorus pentasulfide [1314-80-3] (1992)	1 mg/m³	3 mg/m³	—	222.29	URT irr
Phosphorus trichloride [7719-12-2] (1992)	0.2 ppm	0.5 ppm	—	137.35	URT, eye, & skin irr
Phthalic anhydride [85-44-9] (1992)	1 ppm	—	SEN; A4	148.11	URT, eye, & skin irr
m-Phthalodinitrile [626-17-5] (2008)	5 mg/m³ (IFV)	—	—	128.14	Eye & URT irr
Picloram [1918-02-1] (1992)	10 mg/m³	—	A4	241.48	Liver & kidney dam
Picric acid [88-89-1] (1992)	0.1 mg/m³	—	—	229.11	Skin sens; dermatitis; eye irr
Pindone [83-26-1] (1992)	0.1 mg/m³	—	—	230.25	Coagulation
‡ (Piperazine dihydrochloride) [142-64-3] (1992)	(5 mg/m³)	(—)	(—)	(159.05)	(Eye & skin irr; skin sens; asthma)
Platinum [7440-06-4] (1979)					
Metal	1 mg/m³	—	—	195.09	Asthma; URT irr
Soluble salts, as Pt	0.002 mg/m³	—	—	Varies	Asthma; URT irr
Polyvinyl chloride (PVC) [9002-86-2] (2007)	1 mg/m³ (R)	—	A4	Varies	Pneumoconiosis; LRT irr; pulm func changes

| Substance [CAS No.] (Documentation date) | ADOPTED VALUES | | | MW | TLV® Basis |
	TWA	STEL	Notations		
* Portland cement [65997-15-1] (2009)	1 mg/m³ (E, R)	—	A4	—	Pulm func; resp symptoms; asthma
Potassium hydroxide [1310-58-3] (1992)	—	C 2 mg/m³	—	56.10	URT, eye, & skin irr
Propane [74-98-6]	See Aliphatic hydrocarbon gases: Alkanes [C₁–C₄]				
Propane sultone [1120-71-4] (1976)	—(L)	—	A3	122.14	Cancer
n-Propanol (n-Propyl alcohol) [71-23-8] (2006)	100 ppm	—	A4	60.09	Eye & URT irr
2-Propanol [67-63-0] (2001)	200 ppm	400 ppm	A4	60.09	Eye & URT irr; CNS impair
Propargyl alcohol [107-19-7] (1992)	1 ppm	—	Skin	56.06	Eye irr; liver & kidney dam
β-Propiolactone [57-57-8] (1992)	0.5 ppm	—	A3	72.06	Skin cancer; URT irr
Propionaldehyde [123-38-6] (1998)	20 ppm	—	—	58.1	URT irr
Propionic acid [79-09-4] (1977)	10 ppm	—	—	74.08	Eye, skin, & URT irr
Propoxur [114-26-1] (1992)	0.5 mg/m³	—	A3; BEIₐ	209.24	Cholinesterase inhib
n-Propyl acetate [109-60-4] (1962)	200 ppm	250 ppm	—	102.13	Eye & URT irr
Propylene [115-07-1] (2005)	500 ppm	—	A4	42.08	Asphyxia; URT irr
Propylene dichloride [78-87-5] (2005)	10 ppm	—	SEN; A4	112.99	URT irr; body weight eff
Propylene glycol dinitrate [6423-43-4] (1980)	0.05 ppm	—	Skin; BEIₘ	166.09	Headache; CNS impair

TLV®–CS

	ADOPTED VALUES				
Substance [CAS No.] (*Documentation date*)	TWA	STEL	Notations	MW	TLV® Basis
Propylene oxide [75-56-9] (2000)	2 ppm	—	SEN; A3	58.08	Eye & URT irr
Propyleneimine [75-55-8] (2008)	0.2 ppm	0.4 ppm	Skin; A3	57.09	URT irr; kidney dam
n-Propyl nitrate [627-13-4] (1962)	25 ppm	40 ppm	BEI$_M$	105.09	Nausea; headache
Pyrethrum [8003-34-7] (1992)	5 mg/m^3	—	A4	345 (avg.)	Liver dam; LRT irr
Pyridine [110-86-1] (1992)	1 ppm	—	A3	79.10	Skin irr; liver & kidney dam
Quinone [106-51-4] (1970)	0.1 ppm	—	—	108.09	Eye irr; skin dam
Resorcinol [108-46-3] (1992)	10 ppm	20 ppm	A4	110.11	Eye & skin irr
Rhodium [7440-16-6], as Rh (1981)				102.91	
Metal and insoluble compounds	1 mg/m^3	—	A4	Varies	Metal = URT irr; Insoluble = LRT irr
Soluble compounds	0.01 mg/m^3	—	A4	Varies	Asthma
Ronnel [299-84-3] (2005)	5 mg/m^3 (IFV)	—	A4; BEI$_A$	321.57	Cholinesterase inhib
Rosin core solder thermal decomposition products (colophony) [8050-09-7] (1992)	—(L)	—	SEN	NA	Skin sens; dermatitis; asthma
Rotenone (commercial) [83-79-4] (1992)	5 mg/m^3	—	A4	391.41	URT & eye irr; CNS impair
Selenium [7782-49-2] and compounds, as Se (1992)	0.2 mg/m^3	—	—	78.96	Eye & URT irr
Selenium hexafluoride [7783-79-1], as Se (1992)	0.05 ppm	—	—	192.96	Pulm edema

TLV®–CS

Substance [CAS No.] (Documentation date)	ADOPTED VALUES			MW	TLV® Basis
	TWA	STEL	Notations		
Sesone [136-78-7] (1992)	10 mg/m³	—	A4	309.13	GI irr
Silica, crystalline — α-quartz [14808-60-7; 1317-95-9] and cristobalite [14464-46-1] (2009)	0.025 mg/m³ (R)	—	A2	60.09	Pulm fibrosis; lung cancer
Silicon carbide [409-21-2] (2002)				40.10	
Nonfibrous	10 mg/m³ (I,E)	—	—		URT irr
	3 mg/m³ (R,E)	—	—		URT irr
Fibrous (including whiskers)	0.1 f/cc (F)	—	A2		Mesothelioma; cancer
Silicon tetrahydride [7803-62-5] (1992)	5 ppm	—	—	32.12	URT & skin irr
Silver [7440-22-4] (1992)				107.87	Argyria
Metal, dust and fume	0.1 mg/m³	—	—		
Soluble compounds, as Ag	0.01 mg/m³	—	—	Varies	
‡ (Soapstone (1992))	(6 mg/m³ (E))	(—)	(—)	(—)	(LRT irr)
	(3 mg/m³ (E,R))	(—)	(—)		
Sodium azide [26628-22-8] (1992)				65.02	Card impair; lung dam
as Sodium azide	—	C 0.29 mg/m³	A4		
as Hydrazoic acid vapor	—	C 0.11 ppm	A4		
Sodium bisulfite [7631-90-5] (1992)	5 mg/m³	—	A4	104.07	Skin, eye, & URT irr
Sodium fluoroacetate [62-74-8] (1992)	0.05 mg/m³	—	Skin	100.02	CNS impair; card impair; nausea

Substance [CAS No.] (*Documentation* date)	ADOPTED VALUES			MW	TLV® Basis
	TWA	STEL	Notations		
Sodium hydroxide [1310-73-2] (1992)	—	C 2 mg/m³	—	40.01	URT, eye, & skin irr
Sodium metabisulfite [7681-57-4] (1992)	5 mg/m³	—	A4	190.13	URT irr
Starch [9005-25-8] (1992)	10 mg/m³	—	A4	—	Dermatitis
Stearates(J) (1985)	10 mg/m³	—	A4	Varies	Eye, skin, & URT irr
Stoddard solvent [8052-41-3] (1980)	100 ppm	—	—	140.00	Eye, skin, & kidney dam; nausea; CNS impair
Strontium chromate [7789-06-2], as Cr (1989)	0.0005 mg/m³	—	A2	203.61	Cancer
Strychnine [57-24-9] (1992)	0.15 mg/m³	—	—	334.40	CNS impair
Styrene, monomer [100-42-5] (1996)	20 ppm	40 ppm	A4; BEI	104.16	CNS impair; URT irr; peripheral neuropathy
Subtilisins [1395-21-7; 9014-01-1], as crystalline active enzyme (1972)	—	C 0.00006 mg/m³	—	—	Asthma; skin, URT, & LRT irr
Sucrose [57-50-1] (1992)	10 mg/m³	—	A4	342.30	Dental erosion
Sulfometuron methyl [74222-97-2] (1991)	5 mg/m³	—	A4	364.38	Hematologic eff
Sulfotepp (TEDP) [3689-24-5] (1993)	0.1 mg/m³ (IFV)	—	Skin; A4; BEI_A	322.30	Cholinesterase inhib
Sulfur dioxide [7446-09-5] (2008)	—	0.25 ppm	A4	64.07	Pulm func; LRT irr

| Substance [CAS No.] (Documentation date) | ADOPTED VALUES | | | MW | TLV® Basis |
	TWA	STEL	Notations		
Sulfur hexafluoride [2551-62-4] (1985)	1000 ppm	—	—	146.07	Asphyxia
Sulfuric acid [7664-93-9] (2000)	0.2 mg/m³ (T)	—	A2 (M)	98.08	Pulm func
Sulfur monochloride [10025-67-9] (1986)	—	C 1 ppm	—	135.03	Eye, skin, & URT irr
Sulfur pentafluoride [5714-22-7] (1962)	—	C 0.01 ppm	—	254.11	URT irr, lung dam
Sulfur tetrafluoride [7783-60-0] (1992)	—	C 0.1 ppm	—	108.07	Eye & URT irr; lung dam
Sulfuryl fluoride [2699-79-8] (1992)	5 ppm	10 ppm	—	102.07	CNS impair
Sulprofos [35400-43-2] (2008)	0.1 mg/m³ (IFV)	—	Skin; A4; BEI$_A$	322.43	Cholinesterase inhib
Synthetic vitreous fibers (1999)					
Continuous filament glass fibers	1 f/cc (F)	—	A4	—	
Continuous filament glass fibers	5 mg/m³ (I)	—	A4	—	URT irr
Glass wool fibers	1 f/cc (F)	—	A3	—	URT irr
Rock wool fibers	1 f/cc (F)	—	A3	—	
Slag wool fibers	1 f/cc (F)	—	A3	—	
Special purpose glass fibers	1 f/cc (F)	—	A3	—	
Refractory ceramic fibers	0.2 f/cc (F)	—	A2	—	Pulm fibrosis; pulm func
2,4,5-T [93-76-5] (1992)	10 mg/m³	—	A4	255.49	PNS impair

Substance [CAS No.] (*Documentation date*)	ADOPTED VALUES TWA	ADOPTED VALUES STEL	Notations	MW	TLV® Basis
Talc [14807-96-6] (2009)					
Containing no asbestos fibers	2 mg/m³ (E,R)	—	A4	—	LRT irr
Containing asbestos fibers	Use asbestos TLV® (K)	—	A1	—	
Tellurium [13494-80-9] and compounds (NOS), as Te, excluding hydrogen telluride (1992)	0.1 mg/m³	—	—	127.60	Halitosis
Tellurium hexafluoride [7783-80-4], as Te (1992)	0.02 ppm	—	—	241.61	LRT irr
Temephos [3383-96-8] (2002)	1 mg/m³ (IFV)	—	Skin; A4; BEI$_A$	466.46	Cholinesterase inhib
Terbufos [13071-79-9] (1999)	0.01 mg/m³ (IFV)	—	Skin; A4; BEI$_A$	288.45	Cholinesterase inhib
Terephthalic acid [100-21-0] (1990)	10 mg/m³	—	—	166.13	
Terphenyls [26140-60-3] (1977)	—	C 5 mg/m³	—	230.31	URT & eye irr
1,1,2,2-Tetrabromoethane [79-27-6] (2005)	0.1 ppm (IFV)	—	—	345.70	Eye & URT irr; pulm edema; liver dam
1,1,1,2-Tetrachloro-2,2-difluoroethane [76-11-9] (2007)	100 ppm	—	—	203.83	Liver & kidney dam; CNS impair
1,1,2,2-Tetrachloro-1,2-difluoroethane [76-12-0] (2007)	50 ppm	—	—	203.83	Liver & kidney dam; CNS impair
1,1,2,2-Tetrachloroethane [79-34-5] (1995)	1 ppm	—	Skin; A3	167.86	Liver dam
Tetrachloroethylene [127-18-4] (1990)	25 ppm	100 ppm	A3; BEI	165.80	CNS impair
Tetrachloronaphthalene [1335-88-2] (1992)	2 mg/m³	—	—	265.96	Liver dam

Substance [CAS No.] (*Documentation date*)	ADOPTED VALUES				
	TWA	STEL	Notations	MW	TLV® Basis
Tetraethyl lead [78-00-2], as Pb (1992)	0.1 mg/m³	—	Skin; A4	323.45	CNS impair
Tetraethyl pyrophosphate (TEPP) [107-49-3] (2006)	0.01 mg/m³ (IFV)	—	Skin; BEI_A	290.20	Cholinesterase inhib
Tetrafluoroethylene [116-14-3] (1997)	2 ppm	—	A3	100.20	Kidney & liver dam; liver & kidney cancer
Tetrahydrofuran [109-99-9] (2002)	50 ppm	100 ppm	Skin; A3	72.10	URT irr; CNS impair; kidney dam
Tetrakis (hydroxymethyl) phosphonium salts (2002) Tetrakis (hydroxymethyl) phosphonium chloride [124-64-1]	2 mg/m³		A4	190.56	Body weight; CNS; hepatic
Tetrakis (hydroxymethyl) phosphonium sulfate [55566-30-8]	2 mg/m³	—	SEN; A4	406.26	
Tetramethyl lead [75-74-1], as Pb (1992)	0.15 mg/m³	—	Skin	267.33	CNS impair
Tetramethyl succinonitrile [3333-52-6] (1992)	0.5 ppm	—	Skin	136.20	Headache; nausea; CNS convul
Tetranitromethane [509-14-8] (1992)	0.005 ppm	—	A3	196.04	Eye & URT irr; URT cancer
Tetryl [479-45-8] (1984)	1.5 mg/m³	—	—	287.15	URT irr
* Thallium [7440-28-0] and compounds, as Tl (2009)	0.02 mg/m³ (I)	—	Skin	204.37 Varies	GI dam; peripheral neuropathy
‡ 4,4'-Thiobis(6-tert-butyl-m-cresol) [96-69-5] (1992)	(10 mg/m³)	—	A4	358.52	(Liver & kidney dam)

Substance [CAS No.] (*Documentation* date)	ADOPTED VALUES			MW	TLV® Basis
	TWA	STEL	Notations		
Thioglycolic acid [68-11-1] (1992)	1 ppm	—	Skin	92.12	Eye & skin irr
* Thionyl chloride [7719-09-7] (2009)	—	C 0.2 ppm	—	118.98	URT irr
Thiram [137-26-8] (2007)	0.05 mg/m³ (IFV)	—	SEN; A4	240.44	Body weight & hematologic eff
Tin [7440-31-5], as Sn (1992)					
Metal	2 mg/m³	—	—	118.69	Pneumoconiosis; eye & URT irr; headache; nausea
Oxide & inorganic compounds, except tin hydride	2 mg/m³	—	—	Varies	
Organic compounds	0.1 mg/m³	0.2 mg/m³	Skin; A4	Varies	
Titanium dioxide [13463-67-7] (1992)	10 mg/m³	—	A4	79.90	LRT irr
o-Tolidine [119-93-7] (1992)	—	—	Skin; A3	212.28	Eye, bladder, & kidney irr; bladder cancer; MetHb-emia
Toluene [108-88-3] (2006)	20 ppm	—	A4; BEI	92.13	Visual impair; female repro; pregnancy loss
‡ Toluene-2,4- or 2,6-diisocyanate (or as a mixture) [584-84-9; 91-08-7] (1992)	(0.005 ppm)	(0.02 ppm)	( ); SEN; (A4)	174.15	(Resp sens)
o-Toluidine [95-53-4] (1984)	2 ppm	—	Skin; A3; BEI$_M$	107.15	

TLV®-CS

Substance [CAS No.] (Documentation date)	ADOPTED VALUES			MW	TLV® Basis
	TWA	STEL	Notations		
m-Toluidine [108-44-1] (1984)	2 ppm	—	Skin; A4; BEI$_M$	107.15	Eye, bladder, & kidney irr; MeHb-emia
p-Toluidine [106-49-0] (1984)	2 ppm	—	Skin; A3; BEI$_M$	107.15	MeHb-emia
Tributyl phosphate [126-73-8] (1992)	0.2 ppm	—	BEI$_A$	266.32	Nausea; headache; eye & URT irr
Trichloroacetic acid [76-03-9] (1992)	1 ppm	—	A3	163.39	Eye & URT irr
1,2,4-Trichlorobenzene [120-82-1] (1975)	—	C 5 ppm	—	181.46	Eye & URT irr
1,1,2-Trichloroethane [79-00-5] (1992)	10 ppm	—	Skin; A3	133.41	CNS impair; liver dam
Trichloroethylene [79-01-6] (2006)	10 ppm	25 ppm	A2	131.40	CNS impair; cognitive decrements; renal toxicity
Trichlorofluoromethane [75-69-4] (1992)	—	C 1000 ppm	A4	137.38	Card sens
Trichloronaphthalene [1321-65-9] (1970)	5 mg/m³	—	Skin	231.51	Liver dam; chloracne
1,2,3-Trichloropropane [96-18-4] (1992)	10 ppm	—	Skin; A3	147.43	Liver & kidney dam; eye & URT irr
1,1,2-Trichloro-1,2,2-trifluoroethane [76-13-1] (1992)	1000 ppm	1250 ppm	A4	187.40	CNS impair
Trichlorphon [52-68-6] (1998)	1 mg/m³ (I)	—	A4; BEI$_A$	257.60	Cholinesterase inhib
Triethanolamine [102-71-6] (1990)	5 mg/m³	—	—	149.22	Eye & skin irr
Triethylamine [121-44-8] (1991)	1 ppm	3 ppm	Skin; A4	101.19	Visual impair

Substance [CAS No.] (*Documentation* date)	ADOPTED VALUES				TLV® Basis
	TWA	STEL	Notations	MW	
Trifluorobromomethane [75-63-8] (1979)	1000 ppm	—	—	148.92	CNS & card impair
1,3,5-Triglycidyl-s-triazinetrione [2451-62-9] (1994)	0.05 mg/m³	—	—	297.25	Male repro dam
Trimellitic anhydride [552-30-7] (2007)	0.0005 mg/m³ (IFV)	—	Skin; SEN	192.12	Resp sens
Trimethylamine [75-50-3] (1990)	5 ppm	15 ppm	—	59.11	URT irr
Trimethyl benzene (mixed isomers) [25551-13-7] (1970)	25 ppm	—	—	120.19	CNS impair; asthma; hematologic eff
Trimethyl phosphite [121-45-9] (1980)	2 ppm	—	—	124.08	Eye irr; cholinesterase inhib
2,4,6-Trinitrotoluene (TNT) [118-96-7] (1984)	0.1 mg/m³	—	Skin; BEI$_M$	227.13	MeHb-emia; liver dam; cataract
Triorthocresyl phosphate [78-30-8] (1992)	0.1 mg/m³	—	Skin; A4; BEI$_A$	368.37	Cholinesterase inhib
Triphenyl phosphate [115-86-6] (1992)	3 mg/m³	—	A4	326.28	Cholinesterase inhib
Tungsten [7440-33-7], as W (1979) Metal and insoluble compounds Soluble compounds	5 mg/m³ 1 mg/m³	10 mg/m³ 3 mg/m³	— —	183.85 Varies Varies	LRT irr CNS impair, pulm fibrosis
Turpentine [8006-64-2] and selected monoterpenes [80-56-8; 127-91-3; 13466-78-9] (2001)	20 ppm	—	SEN; A4	136.00 Varies	URT & skin irr; CNS impair; lung dam
Uranium (natural) [7440-61-1] (1992) Soluble and insoluble compounds, as U	0.2 mg/m³	0.6 mg/m³	A1	238.03 Varies	Kidney dam

| Substance [CAS No.] (Documentation date) | ADOPTED VALUES | | | | TLV® Basis |
	TWA	STEL	Notations	MW	
n-Valeraldehyde [110-62-3] (1984)	50 ppm	—	—	86.13	Eye, skin, & URT irr
Vanadium pentoxide [1314-62-1] as V (2008)	0.05 mg/m³ (I)	—	A3	181.88	URT & LRT irr
Vinyl acetate [108-05-4] (1992)	10 ppm	15 ppm	A3	86.09	URT, eye, & skin irr; CNS impair
Vinyl bromide [593-60-2] (1996)	0.5 ppm	—	A2	106.96	Liver cancer
Vinyl chloride [75-01-4] (1997)	1 ppm	—	A1	62.50	Lung cancer; liver dam
4-Vinyl cyclohexene [100-40-3] (1994)	0.1 ppm	—	A3	108.18	Female & male repro dam
Vinyl cyclohexene dioxide [106-87-6] (1994)	0.1 ppm	—	Skin; A3	140.18	Female & male repro dam
Vinyl fluoride [75-02-5] (1996)	1 ppm	—	A2	46.05	Liver cancer; liver dam
N-Vinyl-2-pyrrolidone [88-12-0] (2000)	0.05 ppm	—	A3	111.16	Liver dam
Vinylidene chloride [75-35-4] (1992)	5 ppm	—	A4	96.95	Liver & kidney dam
Vinylidene fluoride [75-38-7] (1996)	500 ppm	—	A4	64.04	Liver dam
Vinyl toluene [25013-15-4] (1992)	50 ppm	100 ppm	A4	118.18	URT & eye irr
Warfarin [81-81-2] (1992)	0.1 mg/m³	—	—	308.32	Coagulation

ADOPTED VALUES

Substance [CAS No.] (Documentation date)	TWA	STEL	Notations	MW	TLV® Basis—Critical Effect(s)
Wood dusts (2009)				NA	
Western red cedar	0.5 mg/m³ (I)	—	SEN; A4		Asthma
All other species	1 mg/m³ (I)	—	—		Pulm func
Carcinogenicity					
Oak and beech	—	—	A1		
Birch, mahogany, teak, walnut	—	—	A2		
All other wood dusts	—	—	A4		
Xylene [1330-20-7] (o, m & p isomers) [95-47-6; 108-38-3; 106-42-3] (1992)	100 ppm	150 ppm	A4; BEI	106.16	URT & eye irr; CNS impair
m-Xylene α,α'-diamine [1477-55-0] (1992)	—	C 0.1 mg/m³	Skin	136.20	Eye, skin, & GI irr
Xylidine (mixed isomers) [1300-73-8] (1999)	0.5 ppm (IFV)	—	Skin; A3; BEI_M	121.18	Liver dam; MeHb-emia
Yttrium [7440-65-5] and compounds, as Y (1986)	1 mg/m³	—	—	88.91	Pulm fibrosis
Zinc chloride fume [7646-85-7] (1992)	1 mg/m³	2 mg/m³	—	136.29	LRT & URT irr
Zinc chromates [13530-65-9; 11103-86-9; 37300-23-5], as Cr (1992)	0.01 mg/m³	—	A1	Varies	Nasal cancer
Zinc oxide [1314-13-2] (2001)	2 mg/m³ (R)	10 mg/m³ (R)	—	81.37	Metal fume fever
Zirconium [7440-67-7] and compounds, as Zr (1992)	5 mg/m³	10 mg/m³	A4	91.22	

TLV®–CS

2010 NOTICE OF INTENDED CHANGES

These substances, with their corresponding values and notations, comprise those for which (1) a limit is proposed for the first time, (2) a change in the Adopted value is proposed, (3) retention as an NIC is proposed, or (4) withdrawal of the *Documentation* and adopted TLV® is proposed. In each case, the proposals should be considered trial values during the period they are on the NIC. These proposals were ratified by the ACGIH® Board of Directors and will remain on the NIC for approximately one year following this ratification. If the Committee neither finds nor receives any substantive data that change its scientific opinion regarding an NIC TLV®, the Committee may then approve its recommendation to the ACGIH® Board of Directors for adoption. If the Committee finds or receives substantive data that change its scientific opinion regarding an NIC TLV®, the Committee may change its recommendation to the ACGIH® Board of Directors for the matter to be either retained on or withdrawn from the NIC.

Documentation is available for each of these substances and their proposed values.

This notice provides an opportunity for comment on these proposals. Comments or suggestions should be accompanied by substantiating evidence in the form of peer-reviewed literature and forwarded in electronic format to The Science Group, ACGIH®, at science@acgih.org. Please refer to the ACGIH® TLV®/BEI® Development Process on the ACGIH® website (http://www.acgih.org/TLV/DevProcess.htm) for a detailed discussion covering this procedure, methods for input to ACGIH®, and deadline date for receiving comments.

| Substance [CAS No.] | 2010 NOTICE OF INTENDED CHANGES | | | | |
	TWA	STEL	Notations	MW	TLV® Basis
† Acetic anhydride [108-24-7]	1 ppm	3 ppm	A4	102.09	URT irr
† Allyl bromide [106-95-6]	0.1 ppm	—	Skin; A4	120.99	Eye & URT irr
† Allyl chloride [107-05-1]	1 ppm	2 ppm	Skin; A3	76.50	Eye & URT irr; liver and kidney dam

2010 NOTICE OF INTENDED CHANGES

Substance [CAS No.]	TWA	STEL	Notations	MW	TLV® Basis
† Calcium silicate [1344-95-2]				—	Pneumoconiosis; pulm func
Nonfibrous particles	0.5 mg/m³ (E, R)	—	A4		
Fibrous forms	1 f/cc (F)	—	A4		
† Carbon black [1333-86-4]	3 mg/m³ (I)	—	A3	—	Bronchitis
† Ethyl benzene [100-41-4]	20 ppm	—	A3; BEI	106.16	URT irr; kidney dam; cochlear impair
† Maleic anhydride [108-31-6]	0.01 mg/m³ (IFV)	—	SEN; A4	98.06	Resp sens
Manganese [7439-96-5] elemental and inorganic compounds, as Mn	0.2 mg/m³ (I) 0.02 mg/m³ (R)		A4	54.94 varies	CNS impair
† Methylacrylonitrile [126-98-7]	1 ppm	—	Skin; A4	67.09	CNS impair, eye & skin irr
† Methyl isopropyl ketone [563-80-4]	20 ppm	—	—	86.14	Embryo/fetal dam
† 2,4-Pentanedione [123-54-6]	25 ppm	—	Skin	100.12	Neurotoxicity; CNS impair
† Piperazine [110-85-0]	0.1 mg/m³ (IFV)	—	SEN; A4	86.14	Resp sens; asthma; liver & kidney dam
† Piperazine dihydrochloride [142-64-3]	WITHDRAW ADOPTED DOCUMENTATION AND TLV®, SEE NIC ENTRY FOR PIPERAZINE				
† Soapstone	WITHDRAW ADOPTED DOCUMENTATION AND TLV®, SEE DOCUMENTATION FOR TALC				
† 4,4'-Thiobis(6-tert-butyl-m-cresol) [96-69-5]	1 mg/m³ (I)	—	A4	358.54	GI tract irr; URT & LRT irr; liver dam
Toluene-2,4- or 2,6-diisocyanate (or as a mixture) [584-84-9; 91-08-7]	0.001 ppm (IFV)	0.003 ppm (IFV)	Skin; SEN; A3	174.15	Asthma

CHEMICAL SUBSTANCES AND OTHER ISSUES UNDER STUDY

The TLV® Chemical Substances Committee solicits information, especially data, which may assist in its deliberations regarding the following substances and issues. Comments and suggestions, accompanied by substantiating evidence in the form of peer-reviewed literature, should be forwarded in electronic format to The Science Group, ACGIH® at science@acgih.org. In addition, the Committee solicits recommendations for additional substances and issues of concern to the industrial hygiene and occupational health communities. Please refer to the ACGIH® TLV®/BEI® Development Process found on the ACGIH® website for a detailed discussion covering this procedure and methods for input to ACGIH® (http://www.acgih.org/ TLV/DevProcess.htm).

The Under Study list is published each year by February 1 on the ACGIH® website (www.acgih.org/TLV/Studies.htm), in the ACGIH® Annual Reports, and later in the annual *TLVs® and BEIs®* book. In addition, the Under Study list is updated by July 31 into a two-tier list.

- Tier 1 entries indicate which chemical substances and physical agents **may** move forward as an NIC or NIE in the upcoming year, based on their status in the development process.

- Tier 2 consists of those chemical substances and physical agents that **will not** move forward, but will either remain on, or be removed from, the Under Study list for the next year.

This updated list will remain in two tiers for the balance of the year. ACGIH® will continue this practice of updating the Under Study list by February 1 and establishing the two-tier list by July 31 each year.

The substances and issues listed below are as of January 1, 2010. *After this date, please refer to the ACGIH® website (http://www.acgih.org/TLV/Studies.htm) for the up-to-date list.*

Chemical Substances

Acetaldehyde
Acetone
Aliphatic hydrocarbon gases
tert-Amyl methyl ether
Atrazine (and related symmetrical triazines)
Barium sulfate
Benz[a]anthracene
Benzidine
Benzo[b]fluoranthene
Benzo[a]pyrene
Bisphenol A
Boron tribromide
Boron trifluoride
Bromodichloromethane
1-Bromopropane
Butane
sec-Butyl acetate
tert-Butyl hydroperoxide
N-Butyl isocyanate
Butylated hydroxytoluene [BHT]
Calcium silicate, synthetic nonfibrous
Carbonyl sulfide
Chlorine
Chrysene
Clopidol
Coal tar pitch volatiles
Cobalt
Creosote
Cyanogen chloride
2,4-D
Diacetyl
Dibutyl phthalate

3,3'-Dichlorobenzidine
1,3-Dichloro-5,5-dimethyl hydantoin
Dicyclopentadiene
Diethyl phthalate
Diethylene glycol monobutyl ether
Diethylhydroxylamine [DEHA]
Diisobutyl ketone
N,N-Dimethyl acetamide
Dimethylformamide
Dipropyl ketone
Ethane
Ethyl cyanoacrylate
Ethyl formate
Ethyl isocyanate
Ethylene oxide
Ethylidene norbornene
Gasoline, all formulations
Glycerin, mist
Hexamethylene diisocyanate
Iodoform
Isobutane
Isophorone diisocyanate
d-Limonene
Liquefied petroleum gas
Lithium hydride
Lithium hydroxide
Mercury, alkyl compounds
Methane
Methomyl
Methyl acetate
Methyl formate
Methyl isoamyl ketone
Methyl isocyanate
Methylene bis (4-cyclohexyliso-
cyanate)
Methylene bisphenyl isocyanate [MDI]
Naphthalene
1-Naphthylamine
2-Naphthylamine
Natural gas
Nickel carbonyl
Nitrogen dioxide
n-Nonane
Nonane, all isomers
Paraquat
Pentachlorophenol
Phenyl isocyanate
Phthalic anhydride
o-Phthalodinitrile
Polycyclic aromatic hydrocarbons [PAHs]
Polymeric MDI
Propane
Simazine
Stearates
Stoddard solvent
Terephthalic acid
Tetramethyl succinitrile
Thioglycolic acid
Titanium dioxide
Tributyl phosphate
Trichloroacetic acid
1,2,3-Trichloropropane
Triethanolamine
Tungsten carbide
Vinyl acetate
5-Vinyl-2-norbornene

Other Issues

1. Definitions of various notations

DEFINITIONS AND NOTATIONS

Definitions

Documentation

The source publication that provides the critical evaluation of the pertinent scientific information and data with reference to literature sources upon which each TLV® or BEI® is based. See the discussion under "TLV®/BEI® Development Process: An Overview" found at the beginning of this book. The general outline used when preparing the Documentation may be found in the Operations Manual of the Threshold Limit Values for Chemical Substances (TLV®-CS) Committee, accessible online at: www.acgih.org/TLV/OPSManual.pdf.

Minimal Oxygen Content

An oxygen (O_2)-deficient atmosphere is defined as one with an ambient ρO_2 less than 132 torr (NIOSH, 1980). The minimum requirement of 19.5% oxygen at sea level (148 torr O_2, dry air) provides an adequate amount of oxygen for most work assignments and includes a margin of safety (NIOSH, 1987; McManus, 1999). Studies of pulmonary physiology suggest that the above requirements provide an adequate level of oxygen pressure in the lungs (alveolar ρO_2 of 60 torr) (Silverthorn, 2001; Guyton, 1991; NIOSH, 1976).

Some gases and vapors, when present in high concentrations in air, act primarily as simple asphyxiants, without other significant physiologic effects. A simple asphyxiant may not be assigned a TLV® because the limiting factor is the available oxygen. Atmospheres deficient in O_2 do not provide adequate warning and most simple asphyxiants are odorless. Account should be taken of this factor in limiting the concentration of the asphyxiant particularly at elevations greater than 5000 feet where the ρO_2 of the atmosphere is less than 120 torr. Several simple asphyxiants present an explosion hazard. Consult the Documentation for further information on specific simple asphyxiants.

Note: See page 81 for Adopted Appendix F: Minimal Oxygen Content.

Notation

A notation is a designation that appears as a component of the TLV® in which specific information is listed in the column devoted to Notations.

Notice of Intended Change (NIC)

The NIC is a list of actions proposed by the TLV®-CS Committee for the coming year. This Notice provides an opportunity for public comment. Values remain on the NIC for approximately one year after they have been ratified by the ACGIH® Board of Directors. The proposals should be considered trial values during the period they are on the NIC. If the Committee neither finds nor receives any substantive data that change its scientific opinion regarding an NIC TLV®, the Committee may then approve its recommendation to the ACGIH® Board of Directors for adoption. If the Committee finds or receives substantive data that change its scientific opinion regarding an NIC TLV®, the Committee may change its recommendation to the ACGIH® Board of Directors for the matter to be either retained on or withdrawn from the NIC.

Values appearing in parentheses in the Adopted TLV® section are to be used during the period in which a proposed change for that value or notation appears on the NIC.

Particulate Matter/Particle Size

For solid and liquid particulate matter, TLVs® are expressed in terms of "total" particulate matter, except where the terms inhalable, thoracic, or respirable particulate mass are used. The intent of ACGIH® is to replace all "total" particulate TLVs® with inhalable, thoracic, or respirable particulate mass TLVs®. Side-by-side sampling using "total" and inhalable, thoracic, or respirable sampling techniques is encouraged to aid in the replacement of current "total" particulate TLVs®. See Appendix C: Particle Size-Selective Sampling Criteria for Airborne Particulate Matter, for the definitions of inhalable, thoracic, and respirable particulate mass.

Particles (insoluble or poorly soluble) Not Otherwise Specified (PNOS)

There are many insoluble particles of low toxicity for which no TLV® has been established. ACGIH® believes that even biologically inert, insoluble, or poorly soluble particles may have adverse effects and suggests that airborne concentrations should be kept below 3 mg/m^3, respirable particles, and 10 mg/m^3, inhalable particles, until such time as a TLV® is set for a particular substance. A description of the rationale for this recommendation and the criteria for substances to which it pertains are provided in Appendix B.

TLV® Basis

TLVs® are derived from publicly available information summarized in their respective Documentation. Although adherence to the TLV® may prevent several adverse health effects, it is not possible to list all of them in this book. The basis on which the values are established will differ from agent to agent (e.g., protection against impairment of health may be a guiding factor for some, whereas reasonable freedom from irritation, narcosis, nuisance, or other forms of stress may form the basis for others). Health impairments considered include those that shorten life expectancy, adversely affect reproductive function or developmental processes, compromise organ or tissue function, or impair the capability for resisting other toxic substances or disease processes.

The TLV® Basis represents the adverse effect(s) upon which the TLV® is based. The TLV® Basis column in this book is intended to provide a field reference for symptoms of overexposure and as a guide for determining whether components of a mixed exposure should be considered as acting independently or additively. Use of the TLV® Basis column is not a substitute for reading the Documentation. Each Documentation is a critical component for proper use of the TLV(s)® and to understand the TLV® basis. A complete list of the TLV® bases used by the Threshold Limit Values for Chemical Substances Committee may be found in their Operations Manual online at: (http://www.acgih.org/TLV/TLV-CS_Ops_Man_2006-2-9.pdf).

Abbreviations used:

card – cardiac	*impair* – impairment
CNS – central nervous system	*inhib* – inhibition
COHb-emia – carboxyhemoglo-	*irr* – irritation
binemia	*LRT* – lower respiratory tract
convul – convulsion	*MeHb-emia* – methemoglobinemia
dam – damage	*PNS* – peripheral nervous system
eff – effects	*pulm* – pulmonary
form – formation	*repro* – reproductive
func – function	*resp* – respiratory
GI – gastrointestinal	*sens* – sensitization
Hb – hemoglobin	*URT* – upper respiratory tract

Notations/Endnotes

Biological Exposure Indices (BEIs®)

The notation "BEI" is listed in the "Notations" column when a BEI® (or BEIs®) is (are) also recommended for the substance. Three subcategories to the "BEI" notation have been added to help the user identify those substances that would use only the BEI® for Acetylcholinesterase Inhibiting Pesticides or Methemoglobin Inducers. They are as follows:

BEI_A = *See* the BEI® for Acetylcholinesterase inhibiting pesticide

BEI_M = *See* the BEI® for Methemoglobin inducers

BEI_P = *See* the BEI® for Polycyclic aromatic hydrocarbons (PAHs)

Biological monitoring should be instituted for such substances to evaluate the total exposure from all sources, including dermal, ingestion, or non-occupational. See the BEI® section in this book and the *Documentation* of the TLVs® and BEIs® for these substances.

Carcinogenicity

A carcinogen is an agent capable of inducing benign or malignant neoplasms. Evidence of carcinogenicity comes from epidemiology, toxicology, and mechanistic studies. Specific notations (i.e., A1, A2, A3, A4, and A5) are used by ACGIH® to define the categories for carcinogenicity and are listed in the Notations column. *See* Appendix A for these categories and definitions and their relevance to humans in occupational settings.

Inhalable Fraction and Vapor (IFV)

The Inhalable Fraction and Vapor (IFV) endnote is used when a material exerts sufficient vapor pressure such that it may be present in both particle and vapor phases, with each contributing a significant portion of the dose at the TLV–TWA concentration. The ratio of the Saturated Vapor Concentration (SVC) to the TLV–TWA is considered when assigning the IFV endnote. The industrial hygienist should also consider both particle and vapor phases to assess exposures from spraying operations, from processes involving temperature changes that may affect the physical state of matter, when a significant fraction of the vapor is dissolved into or adsorbed onto particles of another substance (such as water-soluble compounds in high humidity environments), and in selecting sampling techniques to collect both states of matter (Perez and Soderholm, 1991).

Sensitization

The designation "SEN" in the "Notations" column refers to the potential for an agent to produce sensitization, as confirmed by human or animal data. The SEN notation *does not imply* that sensitization is the critical effect on which the TLV® is based, nor does it imply that this effect is the sole basis for that agent's TLV®. If sensitization data exist, they are carefully considered when recommending the TLV® for the agent. For those TLVs® that are based upon sensitization, they are meant to protect workers from induction of this effect. These TLVs® are not intended to protect those workers who have already become sensitized.

In the workplace, respiratory, dermal, or conjunctival exposures to sensitizing agents may occur. Similarly, sensitizers may evoke respiratory, dermal, or conjunctival reactions. At this time, the notation does not distinguish between sensitization involving any of these organ systems. The absence of a SEN notation does not signify that the agent lacks the ability to produce sensitization but may reflect the paucity or inconclusiveness of scientific evidence.

Sensitization often occurs via an immunologic mechanism and is not to be confused with other conditions or terminology such as hyperreactivity, susceptibility, or sensitivity. Initially, there may be little or no response to a sensitizing agent. However, after a person is sensitized, subsequent exposure may cause intense responses, even at low exposure concentrations (well below the TLV®). These reactions may be life-threatening and may have an immediate or delayed onset. Workers who have become sensitized to a particular agent may also exhibit cross-reactivity to other agents that have similar chemical structures. A reduction in exposure to the sensitizer and its structural analogs generally reduces the incidence of allergic reactions among sensitized individuals. For some sensitized individuals, complete avoidance of exposure to the sensitizer and structural analogs provides the only means to prevent the specific immune response.

Agents that are potent sensitizers present special problems in the workplace. Respiratory, dermal, and conjunctival exposures should be significantly reduced or eliminated through process control measures and personal protective equipment. Education and training (e.g., review of potential health effects, safe handling procedures, emergency information) are also necessary for those who work with known sensitizing agents.

For additional information regarding the sensitization potential of a particular agent, refer to the TLV® *Documentation* for the specific agent.

Skin

The designation "Skin" in the "Notations" column refers to the potential significant contribution to the overall exposure by the cutaneous route, including mucous membranes and the eyes, by contact with vapors, liquids, and solids. Where dermal application studies have shown absorption that could cause systemic effects following exposure, a Skin notation would be considered. The

Skin notation also alerts the industrial hygienist that overexposure may occur following dermal contact with liquid and aerosols, even when airborne exposures are at or below the TLV®.

A Skin notation is not applied to chemicals that may cause dermal irritation. However, it may accompany a SEN notation for substances that cause respiratory sensitization following dermal exposure. Although not considered when assigning a Skin notation, the industrial hygienist should be aware that there are several factors that may significantly enhance potential skin absorption of a substance that otherwise has low potential for the cutaneous route of entry. Certain vehicles can act as carriers, and when pretreated on the skin or mixed with a substance can promote the transfer of the substance into the skin. In addition, the existence of some dermatologic conditions can also significantly affect the entry of substances through the skin or wound.

While relatively limited quantitative data currently exist with regard to skin absorption of gases, vapors, and liquids by workers, ACGIH® recommends that the integration of data from acute dermal studies and repeated-dose dermal studies in animals and humans, along with the ability of the chemical to be absorbed, be used in deciding on the appropriateness of the Skin notation. In general, available data which suggest that the potential for absorption via the hands and forearms during the workday could be significant, especially for chemicals with lower TLVs®, could justify a Skin notation. From acute animal toxicity data, materials having a relatively low dermal LD_{50} (i.e., 1000 mg/kg of body weight or less) would be given a Skin notation. When chemicals penetrate the skin easily (i.e., higher octanol–water partition coefficients) and where extrapolations of systemic effects from other routes of exposure suggest dermal absorption may be important in the expressed toxicity, a Skin notation would be considered. A Skin notation is not applied to chemicals that cause irritation or corrosive effects in the absence of systemic toxicity.

Substances having a Skin notation and a low TLV® may present special problems for operations involving high airborne concentrations of the material, particularly under conditions where significant areas of the skin are exposed for a long period. Under these conditions, special precautions to significantly reduce or preclude skin contact may be required.

Biological monitoring should be considered to determine the relative contribution to the total dose from exposure via the dermal route. ACGIH® recommends a number of adopted Biological Exposure Indices (BEIs®) that provide an additional tool when assessing the total worker exposure to selected materials. For additional information, refer to *Dermal Absorption* in the "Introduction to the Biological Exposure Indices," *Documentation of the Biological Exposure Indices* (2001), and to Leung and Paustenbach (1994). Other selected readings on skin absorption and the skin notation include Sartorelli (2000), Schneider et al. (2000), Wester and Maibach (2000), Kennedy et al. (1993), Fiserova-Bergerova et al. (1990), and Scansetti et al. (1988).

The use of a Skin notation is intended to alert the reader that air sampling alone is insufficient to quantify exposure accurately and that measures to prevent significant cutaneous absorption may be required.

TLV®-CS

References and Selected Reading

American Conference of Governmental Industrial Hygienists: Dermal absorption. In: Documentation of the Biological Exposure Indices, 7th ed., pp. 21–26. ACGIH®, Cincinnati, OH (2001).

Fiserova-Bergerova V; Pierce JT; Droz PO: Dermal absorption potential of industrial chemicals: Criteria for skin notation. Am J Ind Med 17(5):617–635 (1990).

Guyton AC: Textbook of Medical Physiology, 8th ed. W.B. Sanders Co., Philadelphia, PA (1991).

Kennedy Jr GL; Brock WJ; Banerjee AK: Assignment of skin notation for threshold limit values chemicals based on acute dermal toxicity. Appl Occup Environ Hyg 8(1):26–30 (1993).

Leung H; Paustenbach DJ: Techniques for estimating the percutaneous absorption of chemicals due to occupational and environmental exposure. Appl Occup Environ Hyg 9(3):187–197 (1994).

McManus N: Safety and Health in Confined Spaces. Lewis Publishers, Boca Raton, FL (1999).

NIOSH U.S. National Institute for Occupational Safety and Health: A Guide to Industrial Respiratory Protection, DHEW (NIOSH) Pub. No. 76–198. NIOSH, Cincinnati, OH (1976).

NIOSH U.S. National Institute for Occupational Safety and Health: Working in Confined Spaces. DHHS (NIOSH) Pub. No. 80–106. NIOSH, Cincinnati, OH (1980).

NIOSH U.S. National Institute for Occupational Safety and Health: NIOSH Respirator Decision Logic. DHHS (NIOSH) Pub. No. 87–108. NIOSH, Cincinnati, OH (1987).

Perez C; Soderholm SC: Some chemicals requiring special consideration when deciding whether to sample the particle, vapor, or both phases of an atmosphere. Appl Occup Environ Hyg 6:859–864 (1991).

Sartorelli P: Dermal risk assessment in occupational medicine. Med Lav 91(3):183–191 (2000).

Scansetti G; Piolatto G; Rubino GF: Skin notation in the context of workplace exposure standards. Am J Ind Med 14(6):725–732 (1988).

Schneider T; Cherrie JW; Vermeulen R; Kromhout H: Dermal exposure assessment. Ann Occup Hyg 44(7):493–499 (2000).

Silverthorn DE: Human Physiology: An Integrated Approach, 2nd ed. Prentice-Hall, New Jersey (2001).

Wester RC; Maibach HI: Understanding percutaneous absorption for occupational health and safety. Int J Occup Environ Health 6(2):86–92 (2000).

> All pertinent notes relating to the material in the Chemical Substances section of this book appear in the appendices for this section or on the inside back cover.

ADOPTED APPENDICES

APPENDIX A: Carcinogenicity

ACGIH® has been aware of the increasing public concern over chemicals or industrial processes that cause or contribute to increased risk of cancer in workers. More sophisticated methods of bioassay, as well as the use of sophisticated mathematical models that extrapolate the levels of risk among workers, have led to differing interpretations as to which chemicals or processes should be categorized as human carcinogens and what the maximum exposure levels should be. The goal of the Chemical Substances TLV® Committee has been to synthesize the available information in a manner that will be useful to practicing industrial hygienists, without overburdening them with needless details. The categories for carcinogenicity are:

A1 — *Confirmed Human Carcinogen:* The agent is carcinogenic to humans based on the weight of evidence from epidemiologic studies.

A2 — *Suspected Human Carcinogen:* Human data are accepted as adequate in quality but are conflicting or insufficient to classify the agent as a confirmed human carcinogen; OR, the agent is carcinogenic in experimental animals at dose(s), by route(s) of exposure, at site(s), of histologic type(s), or by mechanism(s) considered relevant to worker exposure. The A2 is used primarily when there is limited evidence of carcinogenicity in humans and sufficient evidence of carcinogenicity in experimental animals with relevance to humans.

A3 — *Confirmed Animal Carcinogen with Unknown Relevance to Humans:* The agent is carcinogenic in experimental animals at a relatively high dose, by route(s) of administration, at site(s), of histologic type(s), or by mechanism(s) that may not be relevant to worker exposure. Available epidemiologic studies do not confirm an increased risk of cancer in exposed humans. Available evidence does not suggest that the agent is likely to cause cancer in humans except under uncommon or unlikely routes or levels of exposure.

A4 — *Not Classifiable as a Human Carcinogen:* Agents which cause concern that they could be carcinogenic for humans but which cannot be assessed conclusively because of a lack of data. *In vitro* or animal studies do not provide indications of carcinogenicity which are sufficient to classify the agent into one of the other categories.

A5 — *Not Suspected as a Human Carcinogen:* The agent is not suspected to be a human carcinogen on the basis of properly conducted epidemiologic studies in humans. These studies have sufficiently long follow-up, reliable exposure histories, sufficiently high dose, and adequate statistical power to conclude that exposure to the agent does not convey a significant risk of cancer to humans; OR, the evidence suggesting a lack of carcinogenicity in experimental animals is supported by mechanistic data.

Substances for which no human or experimental animal carcinogenic data have been reported are assigned no carcinogenicity designation.

Exposures to carcinogens must be kept to a minimum. Workers exposed to A1 carcinogens without a TLV® should be properly equipped to eliminate to the fullest extent possible all exposure to the carcinogen. For A1 carcinogens with a TLV® and for A2 and A3 carcinogens, worker exposure by all routes should be carefully controlled to levels as low as possible below the TLV®. Refer to the "Guidelines for the Classification of Occupational Carcinogens" in the Introduction to the Chemical Substances in the *Documentation of the Threshold Limit Values and Biological Exposure Indices* for a complete description and derivation of these designations.

APPENDIX B: Particles (insoluble or poorly soluble) Not Otherwise Specified [PNOS]

The goal of the TLV®-CS Committee is to recommend TLVs® for all substances for which there is evidence of health effects at airborne concentrations encountered in the workplace. When a sufficient body of evidence exists for a particular substance, a TLV® is established. Thus, by definition the substances covered by this recommendation are those for which little data exist. The recommendation at the end of this Appendix is supplied as a guideline rather than a TLV® because it is not possible to meet the standard level of evidence used to assign a TLV®. In addition, the PNOS TLV® and its predecessors have been misused in the past and applied to any unlisted particles rather than those meeting the criteria listed below. The recommendations in this Appendix apply to particles that:

- Do not have an applicable TLV®;

- Are insoluble or poorly soluble in water (or, preferably, in aqueous lung fluid if data are available); and

- Have low toxicity (i.e., are not cytotoxic, genotoxic, or otherwise chemically reactive with lung tissue, and do not emit ionizing radiation, cause immune sensitization, or cause toxic effects other than by inflammation or the mechanism of "lung overload").

ACGIH® believes that even biologically inert, insoluble, or poorly soluble particles may have adverse effects and recommends that airborne concentrations should be kept below 3 mg/m³, respirable particles, and 10 mg/m³, inhalable particles, until such time as a TLV® is set for a particular substance.

APPENDIX C: Particle Size-Selective Sampling Criteria for Airborne Particulate Matter

For chemical substances present in inhaled air as suspensions of solid particles or droplets, the potential hazard depends on particle size as well as mass concentration because of 1) effects of particle size on the deposition site within the respiratory tract and 2) the tendency for many occupational diseases to be associated with material deposited in particular regions of the respiratory tract.

ACGIH® has recommended particle size-selective TLVs® for crystalline silica for many years in recognition of the well-established association between silicosis and respirable mass concentrations. The TLV®-CS Committee is now re-examining other chemical substances encountered in particle form in occupational environments with the objective of defining: 1) the size-fraction most closely associated for each substance with the health effect of concern and 2) the mass concentration within that size fraction which should represent the TLV®.

The Particle Size-Selective TLVs® (PSS–TLVs) are expressed in three forms:

1. *Inhalable Particulate Matter TLVs®* (IPM–TLVs) for those materials that are hazardous when deposited anywhere in the respiratory tract.
2. *Thoracic Particulate Matter TLVs®* (TPM–TLVs) for those materials that are hazardous when deposited anywhere within the lung airways and the gas-exchange region.
3. *Respirable Particulate Matter TLVs®* (RPM–TLVs) for those materials that are hazardous when deposited in the gas-exchange region.

The three particulate matter fractions described above are defined in quantitative terms in accordance with the following equations:[1–3]

A. IPM fraction consists of those particles that are captured according to the following collection efficiency regardless of sampler orientation with respect to wind direction:

$$IPM (d_{ae}) = 0.5 [1 + \exp(-0.06 d_{ae})]$$
$$\text{for } 0 < d_{ae} \leq 100 \ \mu m$$

where: $IPM (d_{ae})$ = the collection efficiency
d_{ae} = aerodynamic diameter of particle in μm

B. TPM fraction consists of those particles that are captured according to the following collection efficiency:

$$TPM (d_{ae}) = IPM (d_{ae}) [1 - F(x)]$$

where: $F(x)$ = cumulative probability function of the standardized normal variable, x

$$x = \frac{\ln(d_{ae}/\Gamma)}{\ln(\Sigma)}$$

$\ln$ = natural logarithm
Γ = 11.64 μm
Σ = 1.5

C. RPM fraction consists of those particles that are captured according to the following collection efficiency:

$$RPM (d_{ae}) = IPM (d_{ae}) [1 - F(x)]$$

where $F(x)$ = same as above, but with Γ = 4.25 μm and Σ = 1.5

The most significant difference from previous definitions is the increase in the median cut point for a respirable particulate matter sampler from 3.5 μm to

4.0 µm; this is in accord with the International Organization for Standardization/European Standardization Committee (ISO/CEN) protocol.[4,5] At this time, no change is recommended for the measurement of respirable particles using a 10-mm nylon cyclone at a flow rate of 1.7 liters per minute. Two analyses of available data indicate that the flow rate of 1.7 liters per minute allows the 10-mm nylon cyclone to approximate the particulate matter concentration which would be measured by an ideal respirable particulate sampler as defined herein.[6,7]

Collection efficiencies representative of several sizes of particles in each of the respective mass fractions are shown in Tables 1, 2, and 3. *Documentation* for the respective algorithms representative of the three mass fractions is found in the literature.[2–4]

TABLE 1. Inhalable Fraction

Particle Aerodynamic Diameter (µm)	Inhalable Particulate Matter (IPM) Fraction Collected (%)
0	100
1	97
2	94
5	87
10	77
20	65
30	58
40	54.5
50	52.5
100	50

TABLE 2. Thoracic Fraction

Particle Aerodynamic Diameter (µm)	Thoracic Particulate Matter (TPM) Fraction Collected (%)
0	100
2	94
4	89
6	80.5
8	67
10	50
12	35
14	23
16	15
18	9.5
20	6
25	2

TABLE 3. Respirable Fraction

Particle Aerodynamic Diameter (µm)	Respirable Particulate Matter (RPM) Fraction Collected (%)
0	100
1	97
2	91
3	74
4	50
5	30
6	17
7	9
8	5
10	1

TLV®-CS

References

1. American Conference of Governmental Industrial Hygienists: Particle Size-Selective Sampling in the Workplace. ACGIH®, Cincinnati, OH (1985).
2. American Conference of Governmental Industrial Hygienists: Particle Size-Selective Sampling for Particulate Air Contaminants. JH Vincent, Ed. ACGIH®, Cincinnati, OH (1999).
3. Soderholm, SC: Proposed International Conventions for Particle Size-Selective Sampling. Ann. Occup. Hyg. 33:301–320 (1989).
4. International Organization for Standardization (ISO): Air Quality—Particle Size Fraction Definitions for Health-Related Sampling. ISO 7708:1995. ISO, Geneva (1995).
5. European Standardization Committee (CEN): Size Fraction Definitions for Measurement of Airborne Particles. CEN EN481:1993. CEN, Brussels (1993).
6. Bartley, DL: Letter to J. Doull, TLV® Committee, July 9, 1991.
7. Lidén, G; Kenny, LC: Optimization of the Performance of Existing Respirable Dust Samplers. Appl. Occup. Environ. Hyg. 8(4):386–391 (1993).

APPENDIX D: Commercially Important Tree Species Suspected of Inducing Sensitization

Common	Latin
SOFTWOODS	
California redwood	*Sequoia sempervirens*
Eastern white cedar	*Thuja occidentalis*
Pine	*Pinus*
Western red cedar	*Thuja plicata*
HARDWOOD	
Ash	*Fraxinus spp.*
Aspen/Poplar/Cottonwood	*Populus*
Beech	*Fagus*
Oak	*Quercus*

TLV®–CS

TROPICAL WOODS

Abirucana	*Pouteria*
African zebra	*Microberlinia*
Antiaris	*Antiaris africana, Antiaris toxicara*
Cabreuva	*Myrocarpus fastigiatus*
Cedar of Lebanon	*Cedra libani*
Central American walnut	*Juglans olanchana*
Cocabolla	*Dalbergia retusa*
African ebony	*Diospryos crassiflora*
Fernam bouc	*Caesalpinia*
Honduras rosewood	*Dalbergia stevensonii*
Iroko or kambala	*Chlorophora excelsa*
Kejaat	*Pterocarpus angolensis*
Kotibe	*Nesorgordonia papaverifera*
Limba	*Terminalia superba*
Mahogany (African)	*Khaya spp.*
Makore	*Tieghemella heckelii*
Mansonia/Beté	*Mansonia altissima*
Nara	*Pterocarpus indicus*
Obeche/African maple/Samba	*Triplochiton scleroxylon*
Okume	*Aucoumea klaineana*
Palisander/Brazilian rosewood/ Tulip wood/Jakaranda	*Dalbergia nigra*
Pau marfim	*Balfourodendron riedelianum*
Ramin	*Gonystylus bancanus*
Soapbark dust	*Quillaja saponaria*
Spindle tree wood	*Euonymus europaeus*
Tanganyike aningre	

APPENDIX E: Threshold Limit Values for Mixtures

Most threshold limit values are developed for a single chemical substance. However, the work environment is often composed of multiple chemical exposures both simultaneously and sequentially. It is recommended that multiple exposures that comprise such work environments be examined to assure that workers do not experience harmful effects.

There are several possible modes of chemical mixture interaction. Additivity occurs when the combined biological effect of the components is equal to the sum of each of the agents given alone. Synergy occurs where the combined effect is greater than the sum of each agent. Antagonism occurs when the combined effect is less.

The general ACGIH® mixture formula applies to the additive model. It is utilized when additional protection is needed to account for this combined effect.

> **The guidance contained in this Appendix does not apply to substances in mixed phases.**

Application of the Additive Mixture Formula

The "TLV® Basis" column found in the table of Adopted Values lists the adverse effect(s) upon which the TLV® is based. This column is a resource that may help alert the reader to the additive possibilities in a chemical mixture and the need to reduce the combined TLV® of the individual components. Note that the column does not list the deleterious effects of the agent, but rather, lists only the adverse effect(s) upon which the threshold limit was based. The current *Documentation of the TLVs® and BEIs®* should be consulted for toxic effects information, which may be of use when assessing mixture exposures.

When two or more hazardous substances have a similar toxicological effect on the same target organ or system, their combined effect, rather than that of either individually, should be given primary consideration. In the absence of information to the contrary, different substances should be considered as additive where the health effect and target organ or system is the same.

That is, if the sum of

$$\frac{C_1}{T_1} + \frac{C_2}{T_2} + \dots \frac{C_n}{T_n}$$

exceeds unity, the threshold limit of the mixture should be considered as being exceeded (where C_1 indicates the observed atmospheric concentration and T_1 is the corresponding threshold limit; see example). It is essential that the atmosphere is analyzed both qualitatively and quantitatively for each component present in order to evaluate the threshold limit of the mixture.

The additive formula applies to simultaneous exposure for hazardous agents with TWA, STEL, and Ceiling values. The threshold limit value time interval base (TWA, STEL, and Ceiling) should be consistent where possible. When agents with the same toxicological effect do not have a corresponding TLV® type, use of mixed threshold limit value types may be warranted. Table E-1 lists possible combinations of threshold limits for the additive mixture formula. Multiple calculations may be necessary.

Where a substance with a STEL or Ceiling limit is mixed with a substance with a TLV–TWA but no STEL, comparison of the short-term limit with the applicable excursion limit may be appropriate. Excursion limits are defined as a value five times the TLV–TWA limit. The amended formula would be:

TABLE E-1. Possible Combinations of Threshold Limits When Applying the Additive Mixture Formula

Full Shift or Short Term	Agent A	Agent B
Full Shift	TLV–TWA	TLV–TWA
Full Shift	TLV–TWA	TLV–Ceiling
Short Term	TLV–STEL	TLV–STEL
Short Term	TLV–Ceiling	TLV–Ceiling
Short Term	Excursion limits where there is no STEL (5 times TLV–TWA value)	TLV–Ceiling or TLV–STEL
Short Term	TLV–STEL	TLV–Ceiling

$$\frac{C_1}{T_{1STEL}} + \frac{C_2}{(T_2)(5)} \leq 1$$

where: T_{1STEL} = the TLV–STEL
T_2 = the TLV–TWA of the agent with no STEL.

The additive model also applies to consecutive exposures of agents that occur during a single work shift. Those substances that have TLV–TWAs (and STELs or excursion limits) should generally be handled the same as if they were the same substance, including attention to the recovery periods for STELs and excursion limits as indicated in the "Introduction to Chemical Substances." The formula does not apply to consecutive exposures of TLV–Ceilings.

Limitations and Special Cases

Exceptions to the above rule may be made when there is a good reason to believe that the chief effects of the different harmful agents are not additive. This can occur when neither the toxicological effect is similar nor the target organ is the same for the components. This can also occur when the mixture interaction causes inhibition of the toxic effect. In such cases, the threshold limit ordinarily is exceeded only when at least one member of the series (C_1/T_1 or C_2/T_2, etc.) itself has a value exceeding unity.

Another exception occurs when mixtures are suspected to have a synergistic effect. The use of the general additive formula may not provide sufficient protection. Such cases at present must be determined individually. Potentiating effects of exposure to such agents by routes other than that of inhalation are also possible. Potentiation is characteristically exhibited at high concentrations, less probably at low. For situations involving synergistic effects, it may be possible to use a modified additive formula that provides additional protection by incorporating a synergy factor. Such treatment of the TLVs® should be used with caution, as the quantitative information concerning synergistic effects is sparse.

Care must be considered for mixtures containing carcinogens in categories A1, A2, or A3. Regardless of application of the mixture formula, exposure to mixtures containing carcinogens should be avoided or maintained as low as possible. See Appendix A.

The additive formula applies to mixtures with a reasonable number of agents. It is not applicable to complex mixtures with many components (e.g., gasoline, diesel exhaust, thermal decomposition products, fly ash, etc.).

Example

A worker's airborne exposure to solvents was monitored for a full shift as well as one short-term exposure. The results are presented in Table E-2.

TABLE E-2. Example Results

Agent	Full-Shift Results (TLV–TWA)	Short-Term Results (TLV–STEL)
1) Acetone	160 ppm	490 ppm
	(500 ppm)	(750 ppm)
2) sec-Butyl acetate	20 ppm	150 ppm
	(200 ppm)	(N/A)
3) Methyl ethyl ketone	90 ppm	220 ppm
	(200 ppm)	(300 ppm)

According to the *Documentation of the TLVs® and BEIs®*, all three substances indicate irritation effects on the respiratory system and thus would be considered additive. Acetone and methyl ethyl ketone exhibit central nervous system effects.

Full shift analysis would utilize the formula:

$$\frac{C_1}{T_1} + \frac{C_2}{T_2} + \frac{C_3}{T_3} \leq 1$$

thus,

$$\frac{160}{500} + \frac{20}{200} + \frac{90}{200} = 0.32 + 0.10 + 0.45 = 0.87$$

The full-shift mixture limit is not exceeded.

Short-term analysis would utilize the formula:

$$\frac{C_1}{T_{1STEL}} + \frac{C_2}{(T_2)(5)} + \frac{C_3}{T_{3STEL}} \leq 1$$

thus,

$$\frac{490}{750} + \frac{150}{1000} + \frac{220}{300} = 0.65 + 0.15 + 0.73 = 1.53$$

The short-term mixture limit is exceeded.

APPENDIX F: Minimal Oxygen Content

Adequate oxygen delivery to the tissues is necessary for sustaining life and depends on 1) the level of oxygen in inspired air, 2) the presence or absence of lung disease, 3) the level of hemoglobin in the blood, 4) the kinetics of oxygen binding to hemoglobin (oxy-hemoglobin dissociation curve), 5) the cardiac output, and 6) local tissue blood flow. For the purpose of the present discussion, only the effects of decreasing the amount of oxygen in inspired air is considered.

The brain and myocardium are the most sensitive tissues to oxygen deficiency. The initial symptoms of oxygen deficiency are increased ventilation, increased cardiac output, and fatigue. Other symptoms that may develop include headache, impaired attention and thought processes, decreased coordination, impaired vision, nausea, unconsciousness, seizures, and death. However, there may be no apparent symptoms prior to unconsciousness. The onset and severity of symptoms depend on many factors such as the magnitude of the oxygen deficiency, duration of exposure, work rate, breathing rate, temperature, health status, age, and pulmonary acclimatization. The initial symptoms of increased breathing and increased heart rate become evident when hemoglobin oxygen saturation is reduced below 90%. At hemoglobin oxygen saturations between 80% and 90%, physiological adjustments occur in healthy adults to resist hypoxia, but in compromised individuals, such as emphysema patients, oxygen therapy would be prescribed for hemoglobin oxygen saturations below 90%. As long as the partial pressure of oxygen (pO_2) in pulmonary capillaries stays above 60 torr, hemoglobin will be more than 90% saturated and normal levels of oxygen transport will be maintained in healthy adults. The alveolar pO_2 level of 60 torr corresponds to 120 torr pO_2 in the ambient air, due to anatomic dead space, carbon dioxide, and water vapor. For additional information on gas exchange and pulmonary physiology see Silverthorn[1] and Guyton.[2]

The U.S. National Institute for Occupational Safety and Health[3] used 60 torr alveolar pO_2 as the physiological limit that establishes an oxygen-deficient atmosphere and has defined an oxygen-deficient atmosphere as one with an ambient pO_2 less than 132 torr.[4] The minimum requirement of 19.5% oxygen at sea level (148 torr pO_2, dry air) provides an adequate amount of oxygen for most work assignments and includes a margin of safety.[5] However, the margin of safety significantly diminishes as the O_2 partial pressure of the atmosphere decreases with increasing altitude, decreases with the passage of low pressure weather events, and decreases with increasing water vapor,[6] such that, at 5000 feet, the pO_2 of the atmosphere may approach 120 torr because of water vapor and the passage of fronts and at elevations greater than 8000 feet, the pO_2 of the atmosphere may be expected to be less than 120 torr.

The physiological effects of oxygen deficiency and oxygen partial pressure variation with altitude for dry air containing 20.948% oxygen are given in Table F-1. No physiological effects due to oxygen deficiency are expected in healthy adults at oxygen partial pressures greater than 132 torr or at elevations less than 5000 feet. Some loss of dark adaptation is reported to occur at elevations greater than 5000 feet. At oxygen partial pressures less than 120 torr (equivalent to an elevation of about 7000 feet or about 5000 feet accounting for water vapor and the passage of low pressure weather events) symptoms in unacclimatized workers include increased pulmonary ventilation and cardiac output, incoordination, and impaired attention and thinking. These symptoms are recognized as being incompatible with safe performance of duties.

Accordingly, ACGIH® recommends a minimal ambient oxygen partial pressure of 132 torr, which is protective against inert oxygen-displacing gases and oxygen-consuming processes for altitudes up to 5000 feet. Figure F-1 is a plot of pO_2 with increasing altitude, showing the recommended minimal value of 132 torr. If the partial pressure of oxygen is less than 132 torr or if it is less than the expected value for that altitude, given in Table F-1, then additional work practices are recommended such as thorough evaluation of the confined space to identify the cause of the low oxygen concentration; use of continuous monitors integrated with warning devices; acclimating workers to the altitude of the work, as adaptation to altitude can increase an individuals work capacity by 70%; use of rest–work cycles with reduced work rates and increased rest periods; training, observation, and monitoring of workers; and easy, rapid access to oxygen-supplying respirators that are properly maintained.

Oxygen-displacing gases may have flammable properties or may produce physiological effects, so that their identity and source should be thoroughly investigated. Some gases and vapors, when present in high concentrations in air, act primarily as simple asphyxiants without other significant physiologic effects. A TLV® may not be recommended for each simple asphyxiant because the limiting factor is the available oxygen. Atmospheres deficient in O_2 do not provide adequate warning and most simple asphyxiants are odorless. Account should be taken of this factor in limiting the concentration of the asphyxiant particularly at elevations greater than 5000 feet where the ρO_2 of the atmosphere may be less than 120 torr.

References

1. Silverthorn DE: Human Physiology: An Integrated Approach, 2nd ed. Prentice-Hall, New Jersey (2001).
2. Guyton AC: Textbook of Medical Physiology, 8th ed. W.B. Saunders Co., Philadelphia (1991).
3. U.S. National Institute for Occupational Safety and Health: A Guide to Industrial Respiratory Protection, DHEW (NIOSH) Pub. No. 76-198. NIOSH, Cincinnati, OH (1976).
4. U.S. National Institute for Occupational Safety and Health: Working in Confined Spaces. DHHS (NIOSH) Pub. No. 80-106. NIOSH, Cincinnati, OH (1979).
5. NIOSH U.S. National Institute for Occupational Safety and Health: NIOSH Respirator Decision Logic. DHHS Pub. No. 87-108. NIOSH, Cincinnati, OH (1987).
6. McManus N: Safety and Health in Confined Spaces. Lewis Publishers, Boca Raton, FL (1999).

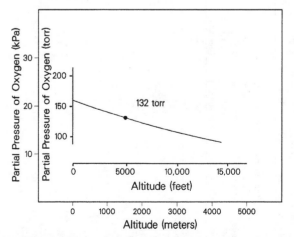

FIGURE F-1. Plot of oxygen partial pressure (ρO_2) (expressed in torr and kPa) with increasing altitude (expressed in feet and meters), showing the recommended oxygen partial pressure of 132 torr.

TABLE F-1. Barometric Pressure, Oxygen Partial Pressure, and Percent Oxygen Concentration Variation with Altitude and Physiological Effect [adapted from McManus[6]]

Altitude Feet (meters)	Barometric Pressure torr, Dry Air[A] (kilopascals)	ρO_2 Equivalent, torr dry air at 20.948% O_2[B] (kilopascals)	%O_2 Equivalent, Dry Air at Sea Level[C] (percent)	Physiological Effect of ρO_2 Levels[D]
0 (0)	760 (101)	159 (21.2)	20.9	
1000 (305)	731 (97.4)	153 (20.4)	20.1	
2000 (610)	704 (93.8)	147 (19.6)	19.3	
3000 (914)	677 (90.3)	142 (18.9)	18.7	
4000 (1219)	652 (86.9)	137 (18.3)	18.0	
5000 (1524)	627 (83.6)	131 (17.5)	17.2	None in healthy adults
6000 (1829)	603 (80.4)	126 (16.8)	16.6	Loss of dark adaptation can occur at elevations above 5000 feet
7000 (2134)	580 (77.3)	121 (16.1)	16.0	Increased pulmonary ventilation and cardiac output, incoordination, and impaired attention and thinking

Altitude feet (meters)	559 (74.5)	117 (15.6)	15.4	Rapid exposure to altitudes over 8000 feet may cause high altitude sickness (respiratory alkalosis, headache, nausea, and vomiting) in unacclimatized individuals. Rapid ascent increases the risk of high altitude pulmonary edema and cerebral edema
8000 (2438)	559 (74.5)	117 (15.6)	15.4	
9000 (2743)	537 (71.6)	112 (14.9)	14.7	
10000 (3048)	517 (68.9)	108 (14.4)	14.2	
11000 (3353)	498 (66.4)	104 (13.9)	13.7	Abnormal fatigue on exertion, faulty coordination, impaired judgment, emotional upset
12000 (3658)	479 (63.8)	100 (13.3)	13.2	
13000 (3962)	461 (61.5)	98 (12.9)	12.8	
14000 (4267)	443 (59.1)	93 (12.4)	12.2	Impaired respiration, very poor judgment and coordination, tunnel vision

[A] Calculated from $P_{re: sea\ level} = 760 \times e^{-(altitude\ in\ feet/25970)}$

[B] Calculated from $pO_2 = 0.20948 \times 760 \times e^{-(altitude\ in\ feet/25970)}$

[C] Calculated from: $P_{\%O_2} = 20.948 \times e^{-(altitude\ in\ feet/25970)}$

[D] The approximate physiological effect in healthy adults is influenced by duration of the oxygen deficiency, work rate, breathing rate, temperature, health status, age and pulmonary acclimatization.

APPENDIX G: Substances Whose Adopted *Documentation* and TLVs® Were Withdrawn For a Variety of Reasons, Including Insufficient Data, Regrouping, Etc.

[Individual entries will remain for a 10-year period, commencing with the year of withdrawal]

Substance [CRN]	Year Withdrawn	Reason
Acetylene tetrabromide	2006	Withdrawn in favor of its IUPAC name; see 1,1,2,2-Tetrabromoethane
Aluminum [7429-90-5] and compounds, as Al	2008	Combined into Aluminum metal and insoluble compounds
Aluminum oxide [1344-28-1]	2008	Combined into Aluminum metal and insoluble compounds
Aluminum welding fumes	2004	TLV® withdrawn as a result of Appendix B removal
APPENDIX B: Substances of Variable Composition	2004	Appendix withdrawn, insufficient data
B1: Polytetrafluoroethylene decomposition products		B1: *Documentation* withdrawn as a result of Appendix removal
B2: Welding fumes (not otherwise specified)		B2: *Documentation* and TLV® withdrawn as a result of Appendix removal
Borates, tetra, sodium salts	2005	Combined into Borate compounds, inorganic
Butane [106-97-8]	2004	Presently covered by Aliphatic hydrocarbon gases: Alkanes [C₁-C₄]
Calcium carbonate [471-34-1]	2007	Insufficient data
Dinitolmide	2007	Withdrawn in favor of its synonym 3,5-Dinitro-o-toluamide
Emery [1302-74-5]	2008	Combined into Aluminum metal and insoluble compounds
Ethane [74-84-0]	2004	Presently covered by Aliphatic hydrocarbon gases: Alkanes [C₁-C₄]
Iron oxide (Fe₂O₃) dust & fume, as Fe	2006	Combined into Iron oxide

APPENDIX G: Substances Whose Adopted *Documentation* and TLVs® Were Withdrawn For a Variety of Reasons, Including Insufficient Data, Regrouping, Etc.

[Individual entries will remain for a 10-year period, commencing with the year of withdrawal] (Con't.)

Substance [CRN]	Year Withdrawn	Reason
Isopropanol	2006	Withdrawn in favor of its IUPAC name, 2-Propanol
Lead arsenate [3687-31-8], as $Pb_3(AsO_4)_2$	2009	Insufficient data
Liquefied petroleum gas (LPG) [68476-85-7]	2004	Insufficient data
Magnesite [546-93-0]	2006	Insufficient data
Methane [74-82-8]	2004	Presently covered by Aliphatic hydrocarbon gases: Alkanes [C_1–C_4]
Oil mist, mineral	2010	Refer to Mineral oil, excluding metal working fluids
Particulates (Insoluble) not otherwise specified	2003	Insufficient data; see Appendix B
Perlite [93763-70-3]	2006	Insufficient data
Propane [74-98-6]	2004	Presently covered by Aliphatic hydrocarbon gases: Alkanes [C_1–C_4]
Rouge	2006	Combined into Iron oxide
Rubber solvent (Naphtha) [8030-30-6]	2009	Refer to Appendix H: Reciprocal Calculation Method for Certain Refined Hydrocarbon Solvent Vapor Mixtures
Silica, amorphous — diatomaceous earth [61790-53-2]	2006	Insufficient data on single-substance exposure, most are co-exposures with crystalline silica
Silica, amorphous — fume [69012-64-2]	2006	Insufficient data

TLV®–CS

APPENDIX G: Substances Whose Adopted *Documentation* and TLVs® Were Withdrawn For a Variety of Reasons, Including Insufficient Data, Regrouping, Etc.

[Individual entries will remain for a 10-year period, commencing with the year of withdrawal] (Con't.)

Substance [CRN]	Year Withdrawn	Reason
Silica, amorphous — fused [60676-86-0]	2006	Insufficient data
Silica amorphous — precipitated silica and silica gel [112926-00-8]	2006	Insufficient data
Silica, crystalline — cristobalite [14464-46-1]	2006	Combined into one TLV® and *Documentation*, i.e., Silica, crystalline
Silica, crystalline — quartz [14808-60-7]	2006	Combined into one TLV® and *Documentation*, i.e., Silica, crystalline
Silica, crystalline — tridymite [15468-32-3]	2005	Insufficient data
Silica, crystalline — tripoli [1317-95-9]	2006	Insufficient data and unlikely single-substance exposure. Combined into one TLV® and *Documentation*, i.e., Silica, crystalline
Silicon [7440-21-3]	2006	Insufficient data
Tantalum [7440-25-7] and Tantalum oxide [1314-61-0] dusts, as Ta	2010	Insufficient data
Tetrasodium pyrophosphate [7722-88-5]	2006	Insufficient data
Triphenyl amine [603-34-9]	2008	Insufficient data
Vegetable oil mist	2006	Insufficient data
VM & P naphtha [8032-32-4]	2009	Refer to Appendix H: Reciprocal Calculation Method for Certain Refined Hydrocarbon Solvent Vapor Mixtures

APPENDIX H: Reciprocal Calculation Method for Certain Refined Hydrocarbon Solvent Vapor Mixtures

The goal of the TLV®-CS Committee is to recommend TLVs® for all substances and mixtures where there is evidence of health effects at airborne concentrations encountered in the workplace. When a sufficient body of evidence exists for a particular substance or mixture, a TLV® is established. However, hydrocarbon solvents are often complex and variable in composition. The use of the mixture formula, found in Appendix E: Threshold Limit Values for Mixtures, is difficult in such cases because these petroleum mixtures contain a large number of unique compounds, many of which do not have a TLV® recommendation.

The reciprocal calculation procedure (RCP) is a method for deriving occupational exposure limits (OEL) for refined hydrocarbon solvents. Refined hydrocarbon solvents often are found as mixtures created by distillation of petroleum oil over a particular boiling range. These mixtures may consist of up to 200 components consisting of aliphatic (alkane), cycloaliphatic (cycloalkane) and aromatic hydrocarbons ranging from 5 to 15 carbons.

There are two aspects of the RCP— the methodology and the group guidance values (GGVs). The methodology is based on the special case formula found in pre-2004 versions of the Mixture Appendix in *TLVs® and BEIs® Based on the Documentation of the Threshold Limit Values for Chemical Substances and Physical Agents and Biological Exposure Indices*. The RCP formula calculates a unique OEL based on the mass composition of the mixture, the GGVs and where applicable, substance-specific TLVs®.

Group guidance values are categorized based on similar chemical and toxicological concerns. Several entities (both trade groups and regulatory authorities) have adopted group guidance values to utilize with the reciprocal mixture formula (RMF) (Farmer, 1995; UK HSE, 2000; McKee et al., 2005). Two examples of published GGVs are found in Table 1. A mixture-specific time-weighted-average limit (GGV-TWA$_{mixture}$) is calculated based on the mass percent makeup of the designated groups utilizing the reciprocal mixture formula and the GGVs from column *B* or *C* and TLV® values in column *D* found in Table 1.

ACGIH® considers this method to be applicable for mixtures if the toxic effects of individual constituents are additive (i.e., similar toxicological effect on the same target organ or system). The principal toxicological effects of hydrocarbon solvent constituents are acute central nervous system (CNS) depression (characterised by effects ranging from dizziness and drowsiness to anaesthesia) and eye and respiratory tract irritation (McKee et al., 2005; ECETOC, 1997).

Application

The RCP is a special use application. It applies only to hydrocarbon solvents containing saturated aliphatics (normal, iso-alkanes and cycloalkanes) and aromatics predominantly consisting of carbon numbers ranging from C_5 to C_{15} derived from petroleum and boiling in the approximate range of 35–320°C. It does not apply to petroleum derived fuels, lubricating oils, or solvent mixtures for which there exists a unique TLV®. It does not apply to hydrocarbons with a toxicity that is significantly greater than the mixture at large, such as benzene (see Limitations below).

Where the mixture is comprised entirely of compounds with unique TLVs®, the mixture should be handled according to Appendix E. When the mixture contains an appreciable amount of a component for which there is a TLV® (i.e., when the use of the TLV® results in a lower GGV-TWA$_{mixture}$), those specific values should be entered into the RCP (see column D, Table 1). When the mixture itself has been assigned a unique TLV®, that value should be utilized rather than the procedures found in this appendix.

Exposure excursions above the calculated GGV-TWA$_{mixture}$ should be handled according to the procedures found in the Introduction to the TLVs® (see Excursion Limits).

The reciprocal calculation mixture formula is:

$$GGV_{mixture} = \frac{1}{\dfrac{F_a}{GGV_a} + ... + \dfrac{F_n}{GGV_n}}$$

where:

$GGV_{mixture}$ = the calculated 8-hour TWA–OEL for the mixture

GGV_a = the guidance value (or TLV®) for group (or component) a

F_a = the liquid mass fraction of group (or component) a in the hydrocarbon mixture (value between 0–1)

GGV_n = the guidance value (or TLV®) for the nth group (or component)

F_n = the liquid mass fraction of the nth group (or component) in the hydrocarbon mixture (value between 0–1)

The resulting GGV$_{mixture}$ should identify the source of GGVs used in the calculation (i.e., column B or C).

The resulting calculated GGV$_{mixture}$ value should follow established recommendations regarding rounding. For calculated values < 100 mg/m^3, round to the nearest 25. For calculated values between 100 and 600 mg/m^3, round to the nearest 50, and for calculated values > 600 mg/m^3, round to the nearest 200 mg/m^3.

Limitations

1. The reciprocal formula requires that the composition of the mixture be characterized at least to the detail of mass percent of the groups found in Table 1.

2. The reciprocal formula does not apply to solvents containing benzene, or n-hexane, or methylnaphthalene, which have individual TLVs® significantly less than the GGV to which they would belong and have unique toxicological properties. Whenever present in the mixture, these components should be measured individually and evaluated using the methodology found in Appendix E, i.e., independent treatment or use of the additive formula depending on the TLV® basis.

3. Care in the use of GGV/RMF should be observed where the mixture in question is known to have significant toxicokinetic interactions of components that are manifested at or below GGV levels.

TABLE 1. Group Guidance Values

A Hydrocarbon Group	B McKee et al. (mg/m³)	C UK–HSE 40/2000 (mg/m³)	D ACGIH® Unique TLVs® (mg/m³)
C5–C6 Alkanes	1500	1800	Pentane, all isomers (1770) Hexane isomers (1760)
C7–C8 Alkanes	1500	1200	Heptane, all isomers (1640) Octane, all isomers (1401)
C5–C6 Cycloalkanes	1500	1800	Cyclopentane (1720) Cyclohexane (350)
C7–C8 Cycloalkanes	1500	800	Methyl cyclohexane (1610)
C7–C8 Aromatics	200	500	Toluene (75) Xylene, all isomers (434) Ethyl benzene (434)
C9–C15 Alkanes	1200	1200	Nonane, all isomers (1050)
C9–C15 Cycloalkanes	1200	800	
C9–C15 Aromatics*	100	500	Trimethyl benzene, isomers (123) Naphthalene (52) Cumen (246)

*n-Hexane (TLV®-176 mg/m³) and methylnaphthalenes (TLV®-3 mg/m³) are significantly below the recommended GGV. Whenever present in the mixture, these components should be measured individually and evaluated using the methodology found in Appendix E, i.e., independent treatment or use of the additive formula depending on the critical effect.

TLV®–CS

4. The use of the reciprocal formula should be restricted to applications where the boiling points of the solvents in the mixture are relatively narrow, within a range of less than 45°C (i.e., vapor pressure within approximately one order of magnitude). The procedure should not be used in situations where the liquid composition is significantly different from the vapor composition. If these conditions cannot be met, the reciprocal formula can be utilized by substituting $F_{(n)}$ in the equation with the vapor mass fraction for each group *(n)* in the hydrocarbon mixture, based on situation-specific airborne concentration measurements.

5. The group guidance values apply only to vapors and do not apply to mists or aerosols. The GGV/RMF procedure does not apply to mixtures containing olefins or other unsaturated compounds or polycyclic aromatic hydrocarbons (PAHs).

Example

A solvent containing the following mass composition is matched with the appropriate group guidance value:

Component	Percent by weight	Group Guidance Value (mg/m³)
C7–C8 alkanes cycloalkanes	45%	1500
C9–C10 alkanes cycloalkanes	40%	1200
C7–C8 aromatics	9%	200
Toluene	6%	75
Benzene	< 1%	-NA-

Based on Column *B*, Table 1 (McKee et al., 2005), the GGV$_{mixture}$ would be:

$$GGV_{mixture} = \frac{1}{\frac{.45}{1500} + \frac{.40}{1200} + \frac{.09}{200} + \frac{.06}{75}} = \frac{1}{.001884}$$

$$= 531 \text{ (rounded to 550 mg/m}^3\text{)}$$

Toluene (part of the aromatic C7, 8 fraction) is added as a TLV® rather than a GGV since it makes a difference in the resulting GGV$_{mixture}$. Benzene would be evaluated separately at the current TLV® for benzene.

References

European Centre for Ecotoxicology and Toxicology of Chemicals (ECETOC). Occupational exposure limits for hydrocarbon solvents. Special Report No. 13. Brussels, Belgium (1997).

Farmer TH: Occupational hygiene limits for hydrocarbon solvents. Annals of Occupational Hygiene 40: 237-242 (1995).

McKee RH; Medeiros AM; Daughtrey WC: A proposed methodology for setting occupational exposure limits for hydrocarbon solvents. J of Occ and Env Hygiene 2: 524-542 (2005).

UK Health and Safety Executive (UKHSE) EH40/2000. Occupational Exposure Limits (2000).

2010
Biological
Exposure
Indices

Adopted by ACGIH®
with Intended Changes

BEIs®

Contents

BEIs®

*Help ensure the continued development of
TLVs® and BEIs®. Make a tax deductible donation to
the FOHS Sustainable TLV®/BEI® Fund today!*

http://www.fohs.org/SusTLV-BEIPrgm.htm

INTRODUCTION TO THE
BIOLOGICAL EXPOSURE INDICES

Biological monitoring provides one means to assess exposure and health risk to workers. It entails measurement of the concentration of a chemical determinant in the biological media of those exposed and is an indicator of the uptake of a substance. Biological Exposure Indices (BEIs®) are guidance values for assessing biological monitoring results. BEIs® represent the levels of determinants that are most likely to be observed in specimens collected from healthy workers who have been exposed to chemicals to the same extent as workers with inhalation exposure at the Threshold Limit Value (TLV®). The exceptions are the BEIs® for chemicals for which the TLVs® are based on protection against nonsystemic effects (e.g., irritation or respiratory impairment) where biological monitoring is desirable because of the potential for significant absorption via an additional route of entry (usually the skin). Biological monitoring indirectly reflects the dose to a worker from exposure to the chemical of interest. The BEI® generally indicates a concentration below which nearly all workers should not experience adverse health effects. The BEI® determinant can be the chemical itself; one or more metabolites; or a characteristic, reversible biochemical change induced by the chemical. In most cases, the specimen used for biological monitoring is urine, blood, or exhaled air. The BEIs® are not intended for use as a measure of adverse effects or for diagnosis of occupational illness.

Biological monitoring can assist the occupational health professional detect and determine absorption via the skin or gastrointestinal system, in addition to that by inhalation; assess body burden; reconstruct past exposure in the absence of other exposure measurements; detect nonoccupational exposure among workers; test the efficacy of personal protective equipment and engineering controls; and monitor work practices.

Biological monitoring serves as a complement to exposure assessment by air sampling. The existence of a BEI® does not indicate a need to conduct biological monitoring. Conducting, designing, and interpreting biological monitoring protocols and the application of the BEI® requires professional experience in occupational health and reference to the current edition of the *Documentation of the Threshold Limit Values and Biological Exposure Indices* (ACGIH®).

DOCUMENTATION

BEIs® are developed by Committee consensus through an analysis and evaluation process. The detailed scientific criteria and justification for each BEI® can be found in the *Documentation of the Threshold Limit Values and Biological Exposure Indices*. The principal material evaluated by the BEI® Committee includes peer-reviewed published data taken from the workplace (i.e., field studies), data from controlled exposure studies, and from appropriate pharmacokinetic modeling when available. The results of animal research are also considered when relevant. The *Documentation* provides essential background information and the scientific reasoning used in establishing each

BEI®. Other information given includes the analytical methods, possible potential for confounding exposures, specimen collection recommendations, limitations, and other pertinent information.

In recommending a BEI®, ACGIH® considers whether published data are of reasonable quality and quantity, and may also consider unpublished data if verified. There are numerous instances when analytical techniques are available for the measurement of a biological determinant, but published information is unavailable or unsuitable for determining a BEI®. In those instances, occupational health professionals are encouraged to accumulate and report biological monitoring data together with exposure and health data.

Relationship of BEIs® to TLVs®

BEI® determinants are an index of an individual's "uptake" of a chemical(s). Air monitoring to determine the TLV® indicates the potential inhalation "exposure" of an individual or group. The uptake within a workgroup may be different for each individual for a variety of reasons, some of which are indicated below. Most BEIs® are based on a direct correlation with the TLV® (i.e., the concentration of the determinant that can be expected when the airborne concentration is at the TLV®). Some of the BEIs® (e.g., lead) are not derived from the TLV®, but directly relate to the development of an adverse health effect. The basis of each BEI® is provided in the *Documentation*.

Inconsistencies may be observed between the information obtained from air monitoring and biological monitoring for a variety of reasons, including, but not limited to, work-related and methodological factors. Examples are listed below:

- Physiological makeup and health status of the worker, such as body build, diet (water and fat intake), metabolism, body fluid composition, age, gender, pregnancy, medication, and disease state.
- Occupational exposure factors, such as the work-rate intensity and duration, skin exposure, temperature and humidity, co-exposure to other chemicals, and other work habits.
- Nonoccupational exposure factors, such as community and home air pollutants, water and food components, personal hygiene, smoking, alcohol and drug intake, exposure to household products, or exposure to chemicals from hobbies or from another workplace.
- Methodological factors, which include specimen contamination or deterioration during collection and storage and bias of the selected analytical method.
- Location of the air monitoring device in relation to the worker's breathing zone.
- Particle size distribution and bioavailability.
- Variable effectiveness of personal protective devices.

Specimen Collection

Because the concentration of some determinants can change rapidly, the specimen collection time (sampling time) is very important and must be observed and recorded carefully. The sampling time is specified in the BEI® and is determined by the duration of retention of the determinant. Substances and determinants that accumulate may not require a specific sampling time. An explanation of the BEI® sampling time is as follows:

Sampling Time	Recommended Collection
1. Prior to shift	16 hours after exposure ceases
2. During shift	Anytime after two hours of exposure
3. End of shift	As soon as possible after exposure ceases
4. End of the workweek	After four or five consecutive working days with exposure
5. Discretionary	At any time

Urine Specimen Acceptability

Urine specimens that are highly dilute or highly concentrated are generally not suitable for monitoring. The World Health Organization has adopted guidelines for acceptable limits on urine specimens as follows:

Creatinine concentration: > 0.3 g/L and < 3.0 g/L
or
Specific gravity: > 1.010 and < 1.030

Specimens falling outside either of these ranges should be discarded and another specimen should be collected. Workers who provide consistently unacceptable urine specimens should be referred for medical evaluation.

Some BEIs® for determinants whose concentration is dependent on urine output are expressed relative to creatinine concentration. For other determinants such as those excreted by diffusion, correction for urine output is not appropriate. In general, the best correction method is chemical-specific, but research data sufficient to identify the best method may not be available. When the field data are only available as adjusted for creatinine, the BEI® will continue to be expressed relative to creatinine; in other circumstances, no correction is recommended, and the BEI® will be expressed as concentration in urine.

Quality Assurance

Each aspect of biological monitoring should be conducted within an effective quality assurance (QA) program. The appropriate specimen must be collected, at the proper time, without contamination or loss, and with use of a suitable container. Donor identification, time of exposure, source of exposure, and the sampling time must be recorded. The analytical method used by the laboratory must have the accuracy, sensitivity, and specificity needed to produce results consistent with the BEI®. Appropriate quality control specimens should be included in the analysis, and the laboratory must follow routine quality control rules. The laboratory should participate in an external proficiency program.

The occupational health professional should provide known blind challenges to the laboratory along with worker specimens (e.g., blanks, purchased or spiked specimens containing the determinant, or split specimens). These blind challenges will enable the occupational health professional to assess the ability of the laboratory to process, analyze, and report results properly, and to have confidence in the laboratory's ability to accurately measure the worker's BEI®. When blind challenges are used, the spiked determinant should be in the same chemical form and matrix as that being analyzed by the laboratory.

Notations

"B" = Background

The determinant may be present in biological specimens collected from subjects who have not been occupationally exposed, at a concentration which could affect interpretation of the result. Such background concentrations are incorporated in the BEI® value.

"Nq" = Nonquantitative

Biological monitoring should be considered for this compound based on the review; however, a specific BEI® could not be determined due to insufficient data.

"Ns" = Nonspecific

The determinant is nonspecific, since it is also observed after exposure to other chemicals.

"Sq" = Semi-quantitative

The biological determinant is an indicator of exposure to the chemical, but the quantitative interpretation of the measurement is ambiguous. These determinants should be used as a screening test if a quantitative test is not practical, or as a confirmatory test if the quantitative test is not specific and the origin of the determinant is in question.

Note:

It is essential to consult the specific BEI® *Documentation* before designing biological monitoring protocols and interpreting BEIs®. In addition, each BEI® *Documentation* now provides a chronology that traces all BEI® recommended actions for the chemical substance in question.

Application of BEIs®

BEIs® are intended as guidelines to be used in the evaluation of potential health hazards in the practice of occupational hygiene. BEIs® do not indicate a sharp distinction between hazardous and nonhazardous exposures. For example, it is possible for an individual's determinant concentration to exceed the BEI® without incurring an increased health risk. If measurements in specimens obtained from a worker on different occasions persistently exceed the BEI®, the cause of the excessive value should be investigated and action taken to reduce the exposure. An investigation is also warranted if the majority of the measurements in specimens obtained from a group of workers at the same workplace and workshift exceed the BEI®. It is desirable that relevant information on related operations in the workplace be recorded.

Due to the variable nature of concentrations in biological specimens, dependence should not be placed on the results of one single specimen. Administrative action should not be normally based on a single isolated measurement, but on measurements of multiple sampling, or an analysis of a repeat specimen. It may be appropriate to remove the worker from exposure following a single high result if there is reason to believe that significant exposure may have occurred. Conversely, observations below the BEI® do not necessarily indicate a lack of health risk.

BEIs® apply to 8-hour exposures, 5 days per week. Although modified work schedules are sometimes used in various occupations, the BEI® Committee does not recommend that any adjustment or correction factor be applied to the BEIs® (i.e., the BEIs® should be used as listed, regardless of the work schedule).

Use of the BEI® should be applied by a knowledgeable occupational health professional. Toxicokinetic and toxicodynamic information is taken into account when establishing the BEI®; thus, some knowledge of the metabolism, distribution, accumulation, excretion, and effect(s) is helpful in using the BEI® effectively. The BEI® is a guideline for the control of potential health hazards to the worker and should not be used for other purposes. The values are inappropriate to use for the general population or for nonoccupational exposures. The BEI® values are neither rigid lines between safe and dangerous concentrations nor are they an index of toxicity.

BEIs®

ADOPTED BIOLOGICAL EXPOSURE DETERMINANTS

Chemical [CAS No.] Determinant	Sampling Time	BE®	Notation
ACETONE [67-64-1]			
Acetone in urine	End of shift	50 mg/L	Ns
ACETYLCHOLINESTERASE INHIBITING PESTICIDES			
Cholinesterase activity in red blood cells	Discretionary	70% of individual's baseline	Ns
ANILINE [62-53-3]			
Aniline★ in urine	End of shift	—	Nq
Aniline released from hemoglobin in blood	End of shift	—	Nq
p-Aminophenol★ in urine	End of shift	50 mg/L	B, Ns, Sq
ARSENIC, ELEMENTAL [7440-38-2] AND SOLUBLE INORGANIC COMPOUNDS (excludes gallium arsenide and arsine)			
Inorganic arsenic plus methylated metabolites in urine	End of workweek	35 µg As/L	B
BENZENE [71-43-2]			
S-Phenylmercapturic acid in urine	End of shift	25 µg/g creatinine	B
t,t-Muconic acid in urine	End of shift	500 µg/g creatinine	B
1,3-BUTADIENE [106-99-0]			
1,2 Dihydroxy-4-(N-acetylcysteinyl)-butane in urine	End of shift	2.5 mg/L	B, Sq
Mixture of N-1 and N-2-(hydroxybutenyl)valine hemoglobin (Hb) adducts in blood	Not critical	2.5 pmol/g Hb	Sq

ADOPTED BIOLOGICAL EXPOSURE DETERMINANTS

Chemical [CAS No.] Determinant	Sampling Time	BEI®	Notation
2-BUTOXYETHANOL [111-76-2] Butoxyacetic acid (BAA) in urine★	End of shift	200 mg/g creatinine	—
CADMIUM [7440-43-9] AND INORGANIC COMPOUNDS Cadmium in urine Cadmium in blood	Not critical Not critical	5 µg/g creatinine 5 µg/L	B B
CARBON DISULFIDE [75-15-0] 2-Thioxothiazolidine-4-carboxylic acid (TTCA) in urine	End of shift	0.5 mg/g creatinine	B, Ns
CARBON MONOXIDE [630-08-0] Carboxyhemoglobin in blood Carbon monoxide in end-exhaled air	End of shift End of shift	3.5% of hemoglobin 20 ppm	B, Ns B, Ns
CHLOROBENZENE [108-90-7] 4-Chlorocatechol in urine★ p-Chlorophenol in urine★	End of shift at end of workweek End of shift at end of workweek	100 mg/g creatinine 20 mg/g creatinine	Ns Ns
CHROMIUM (VI), Water-soluble fume Total chromium in urine Total chromium in urine	End of shift at end of workweek Increase during shift	25 µg/L 10 µg/L	— —

BEIs®

BEIs®

ADOPTED BIOLOGICAL EXPOSURE DETERMINANTS

Chemical [CAS No.] Determinant	Sampling Time	BE®	Notation
COBALT [7440-48-4]			
Cobalt in urine	End of shift at end of workweek	15 µg/L	B
Cobalt in blood	End of shift at end of workweek	1 µg/L	B, Sq
CYCLOHEXANOL [108-93-0]			
1,2-Cyclohexanediol★ in urine	End of shift at end of workweek	—	Nq, Ns
Cyclohexanol★ in urine	End of shift	—	Nq, Ns
CYCLOHEXANONE [108-94-1]			
1,2-Cyclohexanediol★ in urine	End of shift at end of workweek	80 mg/L	Ns, Sq
Cyclohexanol★ in urine	End of shift	8 mg/L	Ns, Sq
DICHLOROMETHANE [75-09-2]			
Dichloromethane in urine	End of shift	0.3 mg/L	Sq
N,N-DIMETHYLACETAMIDE [127-19-5]			
N-Methylacetamide in urine	End of shift at end of workweek	30 mg/g creatinine	—
N,N-DIMETHYLFORMAMIDE (DMF) [68-12-2]			
N-Methylformamide in urine	End of shift	15 mg/L	—
N-Acetyl-S-(N-methylcarbamoyl) cysteine in urine	Prior to last shift of workweek	40 mg/L	Sq
2-ETHOXYETHANOL (EGEE) [110-80-5] and 2-ETHOXYETHYL ACETATE (EGEEA) [111-15-9]			
2-Ethoxyacetic acid in urine	End of shift at end of workweek	100 mg/g creatinine	—

ADOPTED BIOLOGICAL EXPOSURE DETERMINANTS

Chemical [CAS No.] Determinant	Sampling Time	BE®	Notation
ETHYL BENZENE [100-41-4]			
Sum of mandelic acid and phenylglyoxylic acid in urine	End of shift at end of workweek	0.7 g/g creatinine	Ns, Sq
Ethyl benzene in end-exhaled air	Not critical	—	Sq
FLUORIDES			
Fluorides in urine	Prior to shift	3 mg/g creatinine	B, Ns
	End of shift	10 mg/g creatinine	B, Ns
FURFURAL [98-01-1]			
Furoic acid in urine★	End of shift	200 mg/L	Ns
n-HEXANE [110-54-3]			
2,5-Hexanedion☆ in urine	End of shift at end of workweek	0.4 mg/L	—
LEAD [7439-92-1] [See Note below]			
Lead in blood	Not critical	30 µg/100 ml	—

Note: Women of child bearing potential, whose blood Pb exceeds 10 µg/dl, are at risk of delivering a child with a blood Pb over the current Centers for Disease Control guideline of 10 µg/dl. If the blood Pb of such children remains elevated, they may be at increased risk of cognitive deficits. The blood Pb of these children should be closely monitored and appropriate steps should be taken to minimize the child's exposure to environmental lead. (CDC: Preventing Lead Poisoning in Young Children, October 1991; See BE® and TLV® *Documentation* for Lead).

MERCURY			
Total inorganic mercury in urine	Prior to shift	35 µg/g creatinine	B
Total inorganic mercury in blood	End of shift at end of workweek	15 µg/L	B

BEIs®

ADOPTED BIOLOGICAL EXPOSURE DETERMINANTS

Chemical [CAS No.] *Determinant*	*Sampling Time*	*BE®*	*Notation*
METHANOL [67-56-1]			
Methanol in urine	End of shift	15 mg/L	B, Ns
METHEMOGLOBIN INDUCERS			
Methemoglobin in blood	During or end of shift	1.5% of hemoglobin	B, Ns, Sq
* **2-METHOXYETHANOL (EGME)** [109-86-4] and **2-METHOXYETHYL ACETATE (EGMEA)** [110-49-6]			
2-Methoxyacetic acid in urine	End of shift at end of workweek	1 mg/g creatinine	—
METHYL n-BUTYL KETONE [591-78-6]			
2,5-Hexanedione*·ª in urine	End of shift at end of workweek	0.4 mg/L	—
METHYL CHLOROFORM [71-55-6]			
Methyl chloroform in end-exhaled air	Prior to last shift of workweek	40 ppm	—
Trichloroacetic acid in urine	End of workweek	10 mg/L	Ns, Sq
Total trichloroethanol in urine	End of shift at end of workweek	30 mg/L	Ns, Sq
Total trichloroethanol in blood	End of shift at end of workweek	1 mg/L	Ns
4,4'-METHYLENE BIS(2-CHLOROANILINE) **[MBOCA]** [101-14-4]			
Total MBOCA in urine	End of shift	—	Nq
METHYL ETHYL KETONE (MEK) [78-93-3]			
MEK in urine	End of shift	2 mg/L	—

ADOPTED BIOLOGICAL EXPOSURE DETERMINANTS

Chemical [CAS No.] Determinant	Sampling Time	BEI®	Notation
* METHYL ISOBUTYL KETONE (MIBK) [108-10-1]			
MIBK in urine	End of shift	1 mg/L	—
N-METHYL-2-PYRROLIDONE [872-50-4]			
5-Hydroxy-N-methyl-2-pyrrolidone in urine	End of shift	100 mg/L	—
NITROBENZENE [98-95-3]			
Total p-nitrophenol in urine	End of shift at end of workweek	5 mg/g creatinine	Ns
Methemoglobin in blood	End of shift	1.5% of hemoglobin	B, Ns, Sq
PARATHION [56-38-2]			
Total p-nitrophenol in urine	End of shift	0.5 mg/g creatinine	Ns
Cholinesterase activity in red cells	Discretionary	70% of individual's baseline	B, Ns, Sq
PENTACHLOROPHENOL (PCP) [87-86-5]			
Total PCP in urine	Prior to last shift of workweek	2 mg/g creatinine	B
Free PCP in plasma	End of shift	5 mg/L	B
PHENOL [108-95-2]			
Phenol in urine★	End of shift	250 mg/g creatinine	B, Ns
POLYCYCLIC AROMATIC HYDROCARBONS (PAHs)			
1-Hydroxypyrene★(1-HP) in urine	End of shift at end of workweek	—	Nq

BEIs®

BEIs®

ADOPTED BIOLOGICAL EXPOSURE DETERMINANTS

Chemical [CAS No.] Determinant	Sampling Time	BE®	Notation
2-PROPANOL [67-63-0]			
Acetone in urine	End of shift at end of workweek	40 mg/L	B, Ns
STYRENE [100-42-5]			
Mandelic acid plus phenylglyoxylic acid in urine	End of shift	400 mg/g creatinine	Ns
Styrene in venous blood	End of shift	0.2 mg/L	Sq
TETRACHLOROETHYLENE [127-18-4]			
Tetrachloroethylene in end-exhaled air	Prior to shift	3 ppm	—
Tetrachloroethylene in blood	Prior to shift	0.5 mg/L	—
TETRAHYDROFURAN [109-99-9]			
Tetrahydrofuran in urine	End of shift	2 mg/L	—
* TOLUENE [108-88-3]			
Toluene in blood	Prior to last shift of workweek	0.02 mg/L	—
Toluene in urine	End of shift	0.03 mg/L	—
o-Cresol in urine★	End of shift	0.3 mg/g creatinine	B
TRICHLOROETHYLENE [79-01-6]			
Trichloroacetic acid in urine	End of shift at end of workweek	15 mg/L	Ns
Trichloroethanol[1*] in blood	End of shift at end of workweek	0.5 mg/L	Ns
Trichloroethylene in blood	End of shift at end of workweek	—	Sq
Trichloroethylene in end-exhaled air	End of shift at end of workweek	—	Sq

ADOPTED BIOLOGICAL EXPOSURE DETERMINANTS

Chemical [CAS No.] Determinant	Sampling Time	BEI®	Notation
* URANIUM [7440-61-1] Uranium in urine	End of shift	200 µg/L	—
XYLENES [95-47-6; 108-38-3; 106-42-3; 1330-20-7] (technical or commercial grade) Methylhippuric acids in urine	End of shift	1.5 g/g creatinine	—

★ With hydrolysis.
☆ Without hydrolysis; n-hexane, methyl n-butyl ketone and trichloroethylene.

BEIs®

BEIs®

2010 NOTICE OF INTENDED CHANGES

These substances, with their corresponding indices, comprise those for which (1) a BEI® is proposed for the first time, (2) a change in the Adopted index is proposed, (3) retention as an NIC is proposed, or (4) withdrawal of the *Documentation* and adopted BEI® is proposed. In each case, the proposals should be considered trial indices during the period they are on the NIC. These proposals were ratified by the ACGIH® Board of Directors and will remain on the NIC for approximately one year following this ratification. If the Committee neither finds nor receives any substantive data that change its scientific opinion regarding an NIC BEI®, the Committee may then approve its recommendation to the ACGIH® Board of Directors for adoption. If the Committee finds or receives substantive data that change its scientific opinion regarding an NIC BEI®, the Committee may change its recommendation to the ACGIH® Board of Directors for the matter to be either retained on or withdrawn from the NIC.

Documentation is available for each of these substances and their proposed values.

This notice provides an opportunity for comment on these proposals. Comments or suggestions should be accompanied by substantiating evidence in the form of peer-reviewed literature and forwarded in electronic format to The Science Group, ACGIH®, at science@acgih.org. Please refer to the ACGIH® TLV®/BEI® Development Process on the ACGIH® website (http://www.acgih.org/TLV/DevProcess.htm) for a detailed discussion covering this procedure, methods for input to ACGIH®, and deadline date for receiving comments.

2010 NOTICE OF INTENDED CHANGES

Sampling Time	BE®	Notation

There are no substances proposed for the 2010 Notice of Intended Changes.

Chemical [CAS No.]

Determinant

CHEMICAL SUBSTANCES AND OTHER ISSUES UNDER STUDY

The BEI® Committee solicits information, especially data, which may assist it in its deliberations regarding the following substances and issues. Comments and suggestions, accompanied by substantiating evidence in the form of peer-reviewed literature, should be forwarded in electronic format to The Science Group, ACGIH® (science@acgih.org). In addition, the Committee solicits recommendations for additional substances and issues of concern to the industrial hygiene and occupational health communities. Please refer to the ACGIH® TLV®/BEI® Development Process found on the ACGIH® website for a detailed discussion covering this procedure and methods for input to ACGIH® (http://www.acgih.org/TLV/DevProcess.htm).

The Under Study list is published each year by February 1 on the ACGIH® website (www.acgih.org/TLV/Studies.htm), in the ACGIH® Annual Reports, and later in the annual *TLVs® and BEIs®* book. In addition, the Under Study list is updated by July 31 into a two-tier list.

- Tier 1 entries indicate which chemical substances and physical agents **may** move forward as an NIC or NIE in the upcoming year, based on their status in the development process.
- Tier 2 consists of those chemical substances and physical agents that **will not** move forward, but will either remain on, or be removed from, the Under Study list for the next year.

This updated list will remain in two tiers for the balance of the year. ACGIH® will continue this practice of updating the Under Study list by February 1 and establishing the two-tier list by July 31 each year.

The substances and issues listed below are as of January 1, 2010. *After this date, please refer to the ACGIH® website* (http://www.acgih.org/TLV/Studies.htm) *for the up-to-date list.*

Chemical Substances

Beryllium
3,3-Dichlorobenzidine
N,N-Dimethylacetamide
Ethyl benzene
Ethylene oxide
Fluorides
Manganese
Mercury

Methyl ethyl ketone
α-Methyl styrene
Naphthalene
Pentachlorophenol
Thallium
Toluene diisocyanate
Vanadium pentoxide
Xylenes

Other Issues

1. Metabolic polymorphisms
2. Consistency of BEI® *Documentation*
3. BEI® Sampling Strategies/Frequency of Sampling

Feasibility Assessments

For the substances listed below, the BEI® Committee has determined that developing a BEI® is not currently feasible owing to inadequate scientific data. However, the Committee believes that these substances may pose important risks to the health of workers, and therefore, it encourages the submission of new data. Field or experimental studies on the relationship between biological indicators and either health risk or environmental exposure are needed for these agents. A brief summary of the current negative feasibility assessment, including data needs, for each of the listed substances is available from The Science Group, ACGIH®.

Substance	Date of Feasibility Assessment
Acrylonitrile	March 1994
Alachlor	September 2009
Aluminum	September 2007
Antimony	November 1996
Beryllium	September 2002
Chlorpyrifos	October 1996
1,4-Dichlorobenzene	March 1994
2,4-Dichlorophenoxy-acetic acid	March 1994
2-Ethyl hexanoic acid	September 2001
Hydrazines	March 1994
Inorganic borates	October 1995
Manganese	April 1995
Methyl tert-butyl ether	October 1993
Methyl n-butyl ketone	October 1995
Methyl formate	September 2005
Nickel	November 1996
Perfluorooctanoic acid (PFOA)	April 2007
Selenium	October 1995
Trimethylbenzene	August 1999
Vanadium pentoxide	September 2009
Vinyl chloride	August 2002

2010

Threshold Limit Values for Physical Agents in the Work Environment

Adopted by ACGIH®
with Intended Changes

Contents

TLV®-PA

*Help ensure the continued development of
TLVs® and BEIs®. Make a tax deductible donation to
the FOHS Sustainable TLV®/BEI® Fund today!*

http://www.fohs.org/SusTLV-BEIPrgm.htm

INTRODUCTION TO THE PHYSICAL AGENTS

This section presents Threshold Limit Values (TLVs®) for occupational exposure to physical agents of acoustic, electromagnetic, ergonomic, mechanical, and thermal nature. As with other TLVs®, those for physical agents provide guidance on the levels of exposure and conditions under which it is believed that nearly all healthy workers may be repeatedly exposed, day after day, without adverse health effects.

The target organs and health effects of these physical agents vary greatly with their nature; thus, TLVs® are not single numbers, but rather integrations of the measured parameters of the agent, its effects on workers, or both. Due to the many types of physical agents, a variety of scientific disciplines, detection techniques, and instrumentation are applied. Therefore, it is especially important that the physical agents TLVs® be applied only by individuals adequately trained and experienced in the corresponding measurement and evaluation techniques. Given the unavoidable complexity of some of these TLVs®, the most current *Documentation* of the TLVs® for Physical Agents must be consulted when they are applied.

Because of wide variations in individual susceptibility, exposure of an individual at, or even below, the TLV® may result in annoyance, aggravation of a pre-existing condition, or occasionally even physiological damage. Certain individuals may also be hypersusceptible or otherwise unusually responsive to some physical agents at the workplace because of a variety of factors such as genetic predisposition, age, personal habits (e.g., smoking, alcohol, or other drugs), medication, or previous or concurrent exposures. Such workers may not be adequately protected from adverse health effects from exposures to certain physical agents at or below the TLVs®. An occupational physician should evaluate the extent to which such workers require additional protection.

TLVs® are based on available information from industrial experience, from experimental human and animal studies, and when possible, from a combination of the three, as cited in their *Documentation*.

Like all TLVs®, these limits are intended for use in the practice of occupational hygiene and should be interpreted and applied only by a person trained in this discipline. They are not intended for use, or for modification for use, 1) in the evaluation or control of the levels of physical agents in the community or 2) as proof or disproof of an existing physical disability.

These values are reviewed annually by ACGIH® for revision or additions as further information becomes available. ACGIH® regularly examines the data related to mutagenicity, cancer, adverse reproductive effects, and other health effects of physical agents. Comments, accompanied by substantive documentation in the form of peer-reviewed literature, are solicited and should be forwarded in electronic format to The Science Group, ACGIH® (science@acgih.org).

ACGIH® disclaims liability with respect to the use of TLVs®.

TLV®-PA

Notice of Intended Changes

Each year, proposed actions for the forthcoming year are issued in the form of a "Notice of Intended Changes" (NIC). These physical agents, with their corresponding values, comprise those for which (1) a limit is proposed for the first time (i.e., NIE), (2) a change in the Adopted Values are proposed, or (3) retention as an NIC is proposed, or (4) withdrawal of the *Documentation* and adopted TLV® is proposed. In each case, the proposals should be considered trial values during the period they are on the NIC/NIE. These proposals are ratified by the ACGIH® Board of Directors and will remain as NICs/NIEs for approximately one year following this ratification. If the Committee neither finds nor receives any substantive data that change its scientific opinion regarding the TLVs® for a NIC/NIE physical agent, the Committee may then approve its recommendation to the ACGIH® Board of Directors for adoption. If the Committee finds or receives substantive data that change its scientific opinion regarding an NIC/NIE TLV®, the Committee may change its recommendation to the ACGIH® Board of Directors for the matter to be either retained on or withdrawn from the NIC.

Documentation is available for each of these physical agents and their proposed values.

This notice provides an opportunity for comment on these proposals. Comments or suggestions should be accompanied by substantiating evidence in the form of peer-reviewed literature and forwarded in electronic format to The Science Group, ACGIH®, at science@acgih.org. Please refer to the ACGIH® TLV®/BEI® Development Process on the ACGIH® website (http://www.acgih.org/TLV/DevProcess.htm) for a detailed discussion covering this procedure, methods for input to ACGIH®, and deadline date for receiving comments.

Definitions

TLV® categories used in this section include the following:
a) Threshold Limit Value–Time Weighted Average (TLV–TWA). The time-weighted average exposure for an 8-hour workday and 40-hour workweek.
b) Threshold Limit Value–Ceiling (TLV–C). Exposure limit that should not be exceeded even instantaneously.

Carcinogenicity

The Threshold Limit Values for Physical Agents (TLV®-PA) Committee will apply, as necessary, the carcinogenicity designations developed by the Threshold Limit Values for Chemical Substances (TLV®-CS) Committee. Refer to Appendix A in the Chemical Substances section of this *TLVs® and BEIs®* book for these classifications.

Physical and Chemical Factors

Combinations of physical factors such as heat, ultraviolet and ionizing radiation, humidity, abnormal pressure (altitude), and the like, as well as the interaction of physical factors with chemical substances in the workplace, may place added stress on the body so that the effects from exposure at a TLV®

may be altered. This stress may act adversely to increase the toxic response to a foreign substance. Although most TLVs® have built-in uncertainty factors to guard against adverse health effects when there are moderate deviations from normal environments, the uncertainty factors for most exposures are not of such a magnitude as to compensate for gross deviations. In such instances, informed professional judgment must be exercised in the proper adjustment of the TLVs®.

ACOUSTIC

INFRASOUND AND LOW-FREQUENCY SOUND

These limits represent sound exposures to which it is believed nearly all workers may be repeatedly exposed without adverse effects that do not involve hearing.

Except for impulsive sound with durations of less than 2 seconds, one-third octave band[1] levels for frequencies between 1 and 80 Hz should not exceed a sound pressure level (SPL) ceiling limit of 145 dB. In addition, the overall unweighted SPL should not exceed a ceiling limit of 150 dB.

There are no time limits for these exposures. However, application of the TLVs® for Noise and Ultrasound, recommended to prevent noise-induced hearing loss, may provide a reduced acceptable level with time. This reduction will depend upon the amount of attenuation allowed for hearing protection.

An alternative but slightly more constrictive criterion, where the peak SPL measured with the linear or unweighted frequency response of a Sound Level Meter does not exceed 145 dB for nonimpulsive events, may be used. When using this criterion, the measurement instrument should conform to ANSI Standard S1.4 and the linear or unweighted response should extend down to at least 2 Hz.

> Note: Low frequency sounds in the chest resonance range from about 50 Hz to 60 Hz can cause whole-body vibration. Such an effect may cause annoyance and discomfort. The SPL of such sound may need to be reduced to a level where the problem disappears.

References

1. American National Standards Institute: Specification for Octave-Band and Fractional-Octave Band Analog and Digital Filters S1.11-1986 (R1998). ANSI, New York (1998).

TLV®–PA

NOISE

These TLVs® refer to sound pressure levels and durations of exposure that represent conditions under which it is believed that nearly all workers may be repeatedly exposed without adverse effect on their ability to hear and understand normal speech. Prior to 1979, the medical profession had defined hearing impairment as an average hearing threshold level in excess of 25 decibels (ANSI S3.6-1996)[1] at 500, 1000, and 2000 hertz (Hz). The limits that are given here have been established to prevent a hearing loss at higher frequencies, such as 3000 Hz and 4000 Hz. The values should be used as guides in the control of noise exposure and, due to individual susceptibility, should not be regarded as fine lines between safe and dangerous levels.

It should be recognized that the application of the TLVs® for noise will not protect all workers from the adverse effects of noise exposure. The TLVs® should protect the median of the population against a noise-induced hearing loss exceeding 2 dB after 40 years of occupational exposure for the average of 0.5, 1, 2, and 3 kHz. A hearing conservation program with all its elements, including audiometric testing, is necessary when workers are exposed to noise at or above the TLVs®.

Continuous or Intermittent Noise

The sound pressure level should be determined by a sound level meter or dosimeter conforming, as a minimum, to the requirements of the American National Standards Institute (ANSI) Specification for Sound Level Meters, S1.4-1983, Type S2A,[2] or ANSI S1.25-1991 Specification for Personal Noise Dosimeters.[3] The measurement device should be set to use the A-weighted network with slow meter response. The duration of exposure should not exceed that shown in Table 1. These values apply to total duration of exposure per working day regardless of whether this is one continuous exposure or a number of short-term exposures.

When the daily noise exposure is composed of two or more periods of noise exposure of different levels, their combined effect should be considered rather than the individual effect of each. If the sum of the following fractions:

$$\frac{C_1}{T_1} + \frac{C_2}{T_2} + \dots \frac{C_n}{T_n}$$

exceeds unity, then the mixed exposure should be considered to exceed the TLV®. C_1 indicates the total duration of exposure at a specific noise level, and T_1 indicates the total duration of exposure permitted at that level. All on-the-job noise exposures of 80 dBA or greater should be used in the above calculations. With sound level meters, this formula should be used for sounds with steady levels of at least 3 seconds. For sounds in which this condition is not met, a dosimeter or an integrating sound level meter must be used. The TLV® is exceeded when the dose is more than 100% as indicated on a dosimeter set with a 3 dB exchange rate and an 8-hour criteria level of 85 dBA.

The TLV® is exceeded on an integrating sound level meter when the average sound level exceeds the values of Table 1.

Impulsive or Impact Noise

By using the instrumentation specified by ANSI S1.4,[2] S1.25,[3] or IEC 804,[4] impulsive or impact noise is automatically included in the noise measurement. The only requirement is a measurement range between 80 and 140 dBA and the pulse range must be at least 63 dB. No exposures of an unprotected ear in excess of a C-weighted peak sound pressure level of 140 dB should be permitted. If instrumentation is not available to measure a C-weighted peak, an unweighted peak measurement below 140 dB may be used to imply that the C-weighted peak is below 140 dB.

TABLE 1 . TLVs® for Noise[A]

	Duration per Day	**Sound Level dBA[B]**
Hours	24	80
	16	82
	8	85
	4	88
	2	91
	1	94
Minutes	30	97
	15	100
	7.50[C]	103
	3.75[C]	106
	1.88[C]	109
	0.94[C]	112
Seconds[C]	28.12	115
	14.06	118
	7.03	121
	3.52	124
	1.76	127
	0.88	130
	0.44	133
	0.22	136
	0.11	139

[A] No exposure to continuous, intermittent, or impact noise in excess of a peak C-weighted level of 140 dB.

[B] Sound level in decibels are measured on a sound level meter, conforming as a minimum to the requirements of the American National Standards Institute Specification for Sound Level Meters, S1.4 (1983)[2] Type S2A, and set to use the A-weighted network with slow meter response.

[C] Limited by the noise source—not by administrative control. It is also recommended that a dosimeter or integrating sound level meter be used for sounds above 120 dB.

Notes:

1. For impulses above a C-weighted peak of 140 dB, hearing protection should be worn. The MIL-STD-1474C[5] provides guidance for those situations in which single protection (plugs or muffs) or double protection (both muffs and plugs) should be worn.

2. Exposure to certain chemicals may also result in hearing loss. In settings where there may be exposures to noise and to carbon monoxide, lead, manganese, styrene, toluene, or xylene, periodic audiograms are advised and should be carefully reviewed. Other substances under investigation for ototoxic effects include arsenic, carbon disulfide, mercury, and trichloroethylene.

3. There is evidence to suggest that noise exposure in excess of a C-weighted, 8-hour TWA of 115 dBC or a peak exposure of 155 dBC to the abdomen of pregnant workers beyond the fifth month of pregnancy may cause hearing loss in the fetus.

4. The sum of the fractions of any one day may exceed unity, provided that the sum of the fractions over a 7-day period is 5 or less and no daily fraction is more than 3.

5. Table 1 is based on daily exposures in which there will be time away from the workplace in which to relax and sleep. This time away from the workplace will allow any small change to the worker's hearing to recover. When the worker, for times greater than 24 hours, is restricted to a space or series of spaces that serve as both a workplace and a place to relax and sleep, then the background level of the spaces used for relaxation and sleep should be 70 dBA or below.

References

1. American National Standards Institute: Specification for Audiometers. ANSI S3.6- 1996. ANSI, New York (1996).
2. American National Standards Institute: Specification for Sound Level Meters. ANSI S1.4-1983 (R1997). ANSI, New York (1997).
3. American National Standards Institute: Specification for Personal Noise Dosimeters. ANSI S1.25-1991. ANSI, New York (1991).
4. International Electrotechnical Commission: Integrating-Averaging Sound Level Meters. IEC 804. IEC, New York (1985).
5. U.S. Department of Defense: Noise Limits for Military Materiel (Metric). MIL-STD-1474C. U.S. DOD, Washington, DC (1991).

ULTRASOUND

These TLVs® represent conditions under which it is believed that nearly all workers may be repeatedly exposed without adverse effect on their ability to hear and understand normal speech. Previous TLVs® for the frequencies 10 kilohertz (kHz) to 20 kHz, set to prevent subjective effects, are referenced in a cautionary note to Table 1. The 8-hour TWA values are an extension of the TLV® for Noise, which is an 8-hour TWA of 85 dBA. The ceiling values may be verified by using a sound level meter with slow detection and ¹/₃ octave bands. The TWA values may be verified by using an integrating sound level meter with ¹/₃ octave bands. All instrumentation should have adequate frequency response and should meet the specifications of ANSI S1.4-1983 (R1997)[1] and IEC 804.[2]

TABLE 1. TLVs® for Ultrasound

| Mid-Frequency of Third-Octave Band (kHz) | One-third Octave-Band Level[3] | | |
| | Measured in Air in dB re: 20 μPa; Head in Air | | Measured in Water in dB re: 1 μPa; Head in Water |
	Ceiling Values	8-Hour TWA	Ceiling Values
10	105A	88A	167
12.5	105A	89A	167
16	105A	92A	167
20	105A	94A	167
25	110B	—	172
31.5	115B	—	177
40	115B	—	177
50	115B	—	177
63	115B	—	177
80	115B	—	177
100	115B	—	177

A Subjective annoyance and discomfort may occur in some individuals at levels between 75 and 105 dB for the frequencies from 10 kHz to 20 kHz especially if they are tonal in nature. Hearing protection or engineering controls may be needed to prevent subjective effects. Tonal sounds in frequencies below 10 kHz might also need to be reduced to 80 dB.

B These values assume that human coupling with water or other substrate exists. These thresholds may be raised by 30 dB when there is no possibility that the ultrasound can couple with the body by touching water or some other medium. [When the ultrasound source directly contacts the body, the values in the table do not apply. The vibration level at the mastoid bone must be used.] Acceleration Values 15 dB above the reference of 1 g rms should be avoided by reduction of exposure or isolation of the body from the coupling source. (g = acceleration due to the force of gravity, 9.80665 meters/second²; rms = root-mean-square).

References

1. American National Standards Institute: Specification for Sound Level Meters. ANSI S1.4-1983 (R1997). ANSI, New York (1997).
2. International Electrotechnical Commission: Integrating-Averaging Sound Level Meters. IEC 804. IEC, New York (1985).
3. American National Standards Institute: Specification for Octave-Band and Fractional-Octave-Band Analog and Digital Filters S1.11-1986 (R1998). ANSI, New York (1998).

TLV®-PA

ELECTROMAGNETIC RADIATION SPECTRUM AND RELATED TLVs®

Region*	Non-ionizing Radiation						Ionizing Radiation
	Sub-Radiofrequency	Radiofrequency	Microwave	Infrared	Light	Ultraviolet	X-ray
Waveband	ELF			IR-C | IR-B | IR-A		UV-A | UV-B | UV-C	
Wavelength	1000 km — 10 km	1 m	1 mm	3 µm | 1.4 µm | 760 nm	400 nm	315 nm | 280 nm | 180 nm	100 nm
Frequency	300 Hz — 30 kHz	300 MHz	300 GHz				
Applicable TLV*	Sub-Radiofrequency	Radiofrequency and Microwave		Light and Near Infrared		Ultraviolet	Ionizing Radiation
				Lasers			

*The boundaries between regions are set by convention and should not be regarded as absolute dividing lines.

ELECTROMAGNETIC RADIATION AND FIELDS

STATIC MAGNETIC FIELDS

These TLVs® refer to static magnetic field flux densities to which it is believed that nearly all workers may be repeatedly exposed day after day without adverse health effects. These values should be used as guides in the control of exposure to static magnetic fields and should not be regarded as fine lines between safe and dangerous levels.

Routine occupational exposures should not exceed 2 tesla (T) in the general workplace environment, but can have ceiling values of 8 T for workers with special training and operating in a controlled workplace environment. Special training involves making workers aware of transient sensory effects that can result from rapid motion in static magnetic fields with flux densities greater than 2 T. A controlled workplace environment is one in which forces exerted by static magnetic fields on metallic objects do not create potentially hazardous projectiles. Exposure of the limbs of workers in the general workplace environment should not exceed 20 T. Workers with implanted ferromagnetic or electronic medical devices should not be exposed to static magnetic fields exceeding 0.5 mT.

These TLVs® are summarized in Table 1.

TABLE 1. TLVs® for Static Magnetic Fields

Exposure	Ceiling Value
Whole body (general workplace)	2 T
Whole body (special worker training and controlled workplace environment)	8 T
Limbs	20 T
Medical device wearers	0.5 mT

TLV®-PA

SUB-RADIOFREQUENCY (30 kHz and below) MAGNETIC FIELDS

These TLVs® refer to the amplitude of the magnetic flux density (B) of sub-radiofrequency (sub-RF) magnetic fields in the frequency range of 30 kilohertz (kHz) and below to which it is believed that nearly all workers may be exposed repeatedly without adverse health effects. The magnetic field strengths in these TLVs® are root-mean-square (rms) values. These values should be used as guides in the control of exposure to sub-radiofrequency magnetic fields and should not be regarded as fine lines between safe and dangerous levels.

Occupational exposures in the extremely-low-frequency (ELF) range from 1 to 300 hertz (Hz) should not exceed the ceiling value given by the equation:

$$B_{TLV} = \frac{60}{f}$$

where: f = the frequency in Hz

B_{TLV} = the magnetic flux density in millitesla (mT).

For frequencies in the range of 300 Hz to 30 kHz (which includes the voice frequency [VF] band from 300 Hz to 3 kHz and the very-low-frequency [VLF] band from 3 to 30 kHz), occupational exposures should not exceed the ceiling value of 0.2 mT.

These ceiling values for frequencies of 300 Hz to 30 kHz are intended for both partial-body and whole-body exposures. For frequencies below 300 Hz, the TLV® for exposure of the extremities can be increased by a factor of 10 for the hands and feet and by a factor of 5 for the arms and legs.

The magnetic flux density of 60 mT/f at 60 Hz corresponds to a maximum permissible flux density of 1 mT. At 30 kHz, the TLV® is 0.2 mT, which corresponds to a magnetic field intensity of 160 amperes per meter (A/m).

Contact currents from touching ungrounded objects that have acquired an induced electrical charge in a strong sub-RF magnetic field should not exceed the following point contact levels to avoid startle responses or severe electrical shocks:

A. 1.0 milliampere (mA) at frequencies from 1 Hz to 2.5 kHz;
B. 0.4f mA at frequencies from 2.5 to 30 kHz, where f is the frequency expressed in kHz.

Notes:

1. These TLVs® are based on an assessment of available data from laboratory research and human exposure studies. Modifications of the TLVs® will be made if warranted by new information. At this time, there is insufficient information on human responses and possible health effects of magnetic fields in the frequency range of 1 Hz to 30 kHz to permit the establishment of a TLV® for time-weighted average exposures.

TLV®-PA

2. For workers wearing cardiac pacemakers, the TLV® may not protect against electromagnetic interference with pacemaker function. Some models of cardiac pacemakers have been shown to be susceptible to interference by power-frequency (50/60 Hz) magnetic flux densities as low as 0.1 mT. It is recommended that, lacking specific information on electromagnetic interference from the manufacturer, the exposure of persons wearing cardiac pace-makers or similar medical electronic devices be maintained at or below 0.1 mT at power frequencies.

TABLE 1. TLVs® for Sub-Radiofrequency(30 kHz and below) Magnetic Fields

Frequency Range	TLV®
1 to 300 Hz	Whole-body exposure: $\dfrac{60}{f^*}$ ceiling value in mT
1 to 300 Hz	Arms and legs: $\dfrac{300}{f^*}$ ceiling value in mT
1 to 300 Hz	Hands and feet: $\dfrac{600}{f^*}$ ceiling value in mT * where: f = frequency in Hz
300 Hz to 30 kHz	Whole-body and partial-body ceiling value: 0.2 mT
1 Hz to 2.5 kHz	Point contact current limit: 1.0 mA
2.5 to 30 kHz	Point contact current limit: 0.4f mA where: f = frequency in kHz

TLV®-PA

* SUB-RADIOFREQUENCY (30 kHz and below) AND STATIC ELECTRIC FIELDS

These TLVs® refer to the maximum unprotected workplace field strengths of sub-radiofrequency electric fields (30 kHz and below) and static electric fields that represent conditions under which it is believed that nearly all workers may be exposed repeatedly without adverse health effects. The electric field intensities in these TLVs® are root-mean-square (rms) values. The values should be used as guides in the control of exposure and, due to individual susceptibility, should not be regarded as a fine line between safe and dangerous levels. The electric field strengths stated in these TLVs® refer to the field levels present in air, away from the surfaces of conductors (where spark discharges and contact currents may pose significant hazards).

Occupational exposures should not exceed a field strength of 25 kilovolts per meter (kV/m) from 0 hertz (Hz) (direct current [DC]) to 220 Hz. For frequencies in the range of 220 Hz to 3 kilohertz (kHz), the ceiling value is given by:

$$E_{TLV} = 5.525 \times 10^6 / f$$

where: f = the frequency in Hz
E_{TLV} = the rms electric field strength in V/m

A rms value of 1842 V/m is the ceiling value for frequencies from 3 to 30 kHz. These ceiling values are intended for both partial-body and whole-body exposures.

Notes:

1. These TLVs® are based on limiting currents on the body surface and induced internal currents to levels below those that are believed to produce adverse health effects. Certain biological effects have been demonstrated in laboratory studies at electric field strengths below those permitted in the TLV®; however, there is no convincing evidence at the present time that occupational exposure to these field levels leads to adverse health effects.

   Modifications of the TLVs® will be made if warranted by new information. At this time, there is insufficient information on human responses and possible health effects of electric fields in the frequency range of 0 to 30 kHz to permit the establishment of a TLV® for time-weighted average exposures.

2. Field strengths greater than approximately 5 to 7 kV/m can produce a wide range of safety hazards such as startle reactions associated with spark discharges and contact currents from ungrounded conductors within the field. In addition, safety hazards associated with combustion, ignition of flammable materials, and electro-explosive devices may exist

when a high-intensity electric field is present. Care should be taken to eliminate ungrounded objects, to ground such objects, or to use insulated gloves when ungrounded objects must be handled. Prudence dictates the use of protective devices (e.g., suits, gloves, and insulation) in all fields exceeding 15 kV/m.

3. For workers with cardiac pacemakers, the TLV® may not protect against electromagnetic interference with pacemaker function. Some models of cardiac pacemakers have been shown to be susceptible to interference by power-frequency (50/60 Hz) electric fields as low as 2 kV/m. It is recommended that, lacking specific information on electromagnetic interference from the manufacturer, the exposure of pacemaker and medical electronic device wearers should be maintained at or below 1 kV/m.

TLV®–PA

* RADIOFREQUENCY AND MICROWAVE RADIATION

These TLVs® refer to radiofrequency (RF) and microwave radiation in the frequency range of 30 kilohertz (kHz) to 300 gigahertz (GHz) and represent conditions under which it is believed nearly all workers may be repeatedly exposed without adverse health effects. The TLVs®, in terms of root-mean-square (rms), electric (E), and magnetic (H) field strengths, the equivalent plane-wave free-space power densities (S), and induced currents (I) in the body that can be associated with exposure to such fields, are given in Table 1 as a function of frequency, f, in megahertz (MHz).

A. The TLVs® in Table 1, Part A, refer to exposure values obtained by spatially averaging over an area equivalent to the vertical cross-section of the human body (projected area). In the case of partial body exposure, the TLVs® can be relaxed. In nonuniform fields, spatial peak values of field strength may exceed the TLVs® if the spatially averaged value remains within the specified limits. The TLVs® may also be relaxed by reference to specific absorption rate (SAR) limits by appropriate calculations or measurements.

B. Access should be restricted to limit the rms RF body current and potential for RF electrostimulation ("shock," below 0.1 MHz) or perceptible heating (at or above 0.1 MHz) as follows (see Table 1, Part B):

1. For freestanding individuals (no contact with metallic objects), RF current induced in the human body, as measured through either foot, should not exceed the following values:

   I = 1000 f mA for (0.03 < f < 0.1 MHz) averaged over 0.2 s, where mA = milliampere; and

   I = 100 mA for (0.1 < f < 100 MHz) averaged over 6 min.

TABLE 1. Radiofrequency and Microwave TLVs®

Part A—Electromagnetic Fields[A] (f = frequency in MHz)			
Frequency	Power Density, S (W/m²)	Electric Field Strength, E (V/m)	Magnetic Field Strength, H (A/m)	Averaging Time E², H², or S (min)
30 kHz–100 kHz		1842	163	6
100 kHz–1 MHz		1842	16.3/f	6
1 MHz–30 MHz		1842/f	16.3/f	6
30 MHz–100 MHz		61.4	16.3/f	6
100 MHz–300 MHz	10	61.4	0.163	6
300 MHz–3 GHz	f/30			6
3 GHz–30 GHz	100			$33,878.2/f^{1.079}$
30 GHz–300 GHz	100			$67.62/f^{0.476}$

[A]The exposure values in terms of electric and magnetic field strengths are obtained by spatially averaging over an area equivalent to the vertical cross-section of the human body (projected area). At frequencies above 30 GHz, the power density TLV® is the limit over any contiguous 0.01 m² of body surface.

- -

Part B—Induced and Contact Radiofrequency Currents[B]
Maximum Current (mA)

Frequency	Through Both Feet	Through Either Foot	Grasping	Averaging Time
30 kHz–100 kHz	2000 f	1000 f	1000 f	0.2 s[C]
100 kHz–100 MHz	200	100	100	6 min[D]

[B] It should be noted that the current limits given above may not adequately protect against startle reactions and burns caused by transient discharges when contacting an energized object. Maximum touch current is limited to 50% of the maximum grasping current. The ceiling value for induced and contact currents is 500 mA.

[C] I is averaged over a 0.2 s period.

[D] I is averaged over a 6-min period (e.g., for either foot or hand contact, i.e., $I t \leq 60,000$ mA^2-min).

2. For conditions of possible contact with metallic bodies, maximum RF current through an impedance equivalent to that of the human body for conditions of grasping contact as measured with a contact current meter should not exceed the following values:

I = 1000 f mA for (0.03 < f < 0.1 MHz) averaged over 0.2 s; and

I = 100 mA for (0.1 < f < 100 MHz) averaged over 6 min.

3. For touch contact with conductive objects, the maximum RF current should not exceed more than one-half the maximum RF current for grasping contact. The means of compliance with these current limits can be determined by the user of the TLVs® as appropriate. The use of protective gloves, the avoidance of touch contact with conductive objects, the prohibition of metallic objects, or training of personnel may be sufficient to ensure compliance with these TLVs®. Evaluation of the magnitude of the induced currents will normally require a direct measurement. However, induced and contact current measurements are not required if the spatially averaged electric field strength does not exceed the TLV® given in Table 1, Part A at frequencies between 0.1 and 100 MHz, as shown graphically in Figure 2.

C. For near-field exposures at frequencies less than 300 MHz, the applicable TLV® is given in terms of rms electric and magnetic field strength, as shown in Table 1, Part A. Equivalent plane-wave power density, S (in W/m²) can be calculated from field strength measurement data as follows:

$$S = \frac{E^2}{377}$$

where: E^2 is in volts squared (V^2) per meter squared (m^2); and

$$S = 377 H^2$$

where: H^2 is in amperes squared (A^2) per meter squared (m^2).

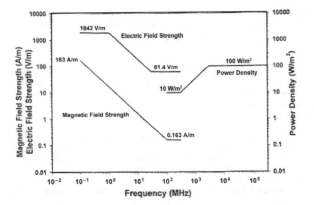

FIGURE 1. Threshold Limit Values (TLVs®) for Radio-frequency/Microwave Radiation in the workplace (for whole-body specific absorption rate [SAR] < 0.4 W/kg. Reprinted with permission of IEEE from Std. C95.1 – 2005.

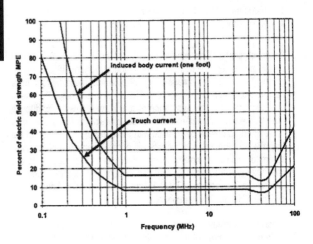

FIGURE 2. Percent of electric field strength TLVs® below which induced and contact current limits are *not* required from 0.1 to 100 MHz. Reprinted with permission of IEEE from Std. C95.1 – 2005.

Figure 3 can assist the user of the TLV® in making E, H, and current measurements in the correct order of precedence.

D. For exposures to pulsed fields of pulse duration less than 100 milliseconds (ms) at frequencies in the range 0.1 MHz to 300 GHz, the maximum value of the instantaneous E field is 100 kV/m. The total incident energy density during any 100 ms period within the averaging time (see Table 1, Part A) shall not exceed 20% of the total specific energy absorption (SA) permitted during the entire averaging time for a continuous field, i.e., $0.2 \times 144 = 28.8$ J/kg. For pulse durations greater than 100 ms, normal time-averaging calculations apply.

The TLV® values in Table 1 should be used as guides in the evaluation and control of exposure to radiofrequency and microwave radiation and should not be regarded as fine lines between safe and dangerous levels. The values of E, H and S given in Table 1, Part A are shown graphically as a function of frequency in Figure 1. Figure 2 depicts the maximum permissible current values given in Table 1, Part B through one foot or touch current as a function of the maximum permissible electric field strength TLV® over the frequency range 0.1 to 100 MHz.

Notes:

1. It is believed that workers may be exposed repeatedly to fields up to these TLVs® without adverse health effects. Nevertheless, personnel should not needlessly be exposed to higher levels of RF radiation, approaching the TLVs®, when simple measures can prevent it.
2. For mixed or broadband fields at a number of frequencies for which there are different values of the TLV® (in terms of E^2, H^2, or S) incurred within each frequency interval should be determined and the sum of all such fractions should not exceed unity.
3. The TLVs® refer to values averaged over any 6-min (0.1-h) period for frequencies less than 3 GHz, and over shorter periods for higher frequencies down to 10 seconds at 300 GHz, as indicated in Table 1, Part A.
4. At frequencies between 0.1 and 3 GHz, the TLVs® for electromagnetic field strengths may be exceeded if:
   a) the exposure conditions can be shown by appropriate techniques to produce SARs below 0.4 W/kg, as averaged over the whole body;
   b) the induced currents in the body conform with the TLVs® in Table 1, Part B; and
   c) spatial peak SAR values do not exceed 10 W/kg, as averaged over any cubic volume with 10 g of tissue, except for the hands, wrists, feet, ankles, and pinnae, where the spatial peak SAR exposure should not exceed 20 W/kg averaged over any cubic volume of tissue containing 10 g. The SARs are to be averaged over 6 min. Recognition should be given to regions of the body where a 10 cm^3 volume may have a mass significantly less than 10 g because of enclosed voids containing air. In these regions the absorbed power should be divided by the actual mass to determine spatial peak SARs.
5. Above 3 GHz, relaxation of the TLV® conditions may be permissible under partial body exposure conditions.

6. The measurement of RF field strength depends upon several factors, including probe dimensions and distance of the source from the probe. Measurement procedures should follow the recommendations given in IEEE C95.3-2002 (IEEE, 2002) and Report No. 119 of the National Council on Radiation Protection and Measurements (NCRP, 1993).

7. All exposures should be limited to a maximum (peak) electric field intensity of 100 kV/m.

8. Ultrawideband (UWB) radiation is a relatively new modality used for imaging, wireless communications (voice, data, and video), identification tags, security systems, and other applications. UWB signals consist of short pulses (usually < 10 nanoseconds [ns]) and fast rise time (< 200 picoseconds [ps]) that result in a very wide bandwidth. For practical purposes, UWB can be considered as a signal that has a bandwidth greater than the central frequency. The following is a set of guidelines for human exposure to UWB radiation that follows the recommendations of the Tri-Service Electromagnetic Radiation Panel approved in May 1996. For a UWB pulse, the specific absorption rate (SAR) expressed in W/kg of tissue is given by:

$$SAR = S \times PW \times PRF \times 0.025$$

where:

S = equivalent plane-wave power density (W/m²);
PW = effective pulse width (s), including the ring-down time;
PRF = pulse repetition frequency (s⁻¹); and
0.025 = maximum normalized SAR (W/kg) per W/m² in the human body exposed to a 70-MHz RF field.

Exposure limitations are considered for two conditions: (A) UWB exposure > 6 min and (B) UWB exposure < 6 min with an SAR > 0.4 W/kg, the whole-body limit allowed by the IEEE C95.1 standard for RF radiation issued in 1991 and revised in 1999 and 2005.

Condition A: For exposures > 6 min, the SAR is limited to 0.4 W/kg, averaged over any 6 min period, corresponding to an SA value of 144 J/kg for 6 min. The permitted PRF for a UWB pulse is given by the following:

$$PRF(s^{-1}) = \frac{144 \text{ J/kg}}{(SA \text{ in J/kg per pulse})(360 \text{ s})}$$

Condition B: The conservative assumption is made that the permissible exposure time (ET) is inversely proportional to the square of the SAR in W²/kg². ET is then given by the following equation:

$$ET(s) = \frac{(0.4 \text{ W/kg} \times 144 \text{ J/kg})}{(SAR)^2} = \frac{57.6}{(SAR)^2}$$

9. Many devices used in medicine, manufacturing, telecommunications, and transportation are highly sensitive to interference by exposure to radiofrequency fields (RFI). This problem has increased as a result of the rapid growth in the use of wireless communication devices, such as cellular telephones, handheld transceivers, and vehicle-mounted transceivers. The U.S. Food and Drug Administration's Center for Devices and Radiological Health has made a major effort to inform manufacturers of the need to

make medical devices immune to RFI effects to the maximum extent possible. However, RFI problems continue to be identified and can adversely affect the operation of cardiac pacemakers, defibrillators, drug infusion pumps, apnea monitors, and a variety of other medical devices such as electrically powered wheel-chairs. For these devices, the TLVs® may not protect against RFI. The use of sensitive medical equipment or the entry of individuals wearing medical electronic devices subject to RFI should be restricted to locations where the levels of RF-microwave fields at frequencies up to 3 GHz are not expected to interfere with operation of medical devices based on manufacturers' specifications (typically field levels below 3 to 10 V/m that meet compliance requirements for immunity to RFI).

References

Institute of Electrical and Electronic Engineers (IEEE): IEEE Recommended Practice for Measurements and Computations of Radiofrequency Electromagnetic Fields with Respect to Human Exposure to Such Fields, 100 kHz – 300 GHz. IEEE C95.3-2002. IEEE, New York (2002).

National Council on Radiation Protection and Measurements: A Practical Guide to the Determination of Human Exposures to Radiofrequency Fields. Report No 119. NCRP, Bethesda, MD (1993).

Tri-Service Electromagnetic Radiation Panel: Ultra-wideband (UWB) Interim Guidance. Approved May 1996. Available from Brooks Air Force Base, San Antonio, Texas.

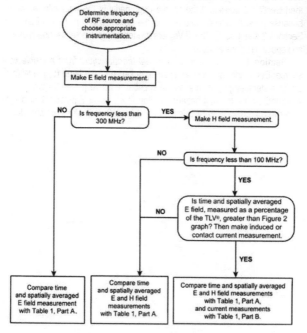

FIGURE 3. Flowchart for making E, H, and current measurements in the correct order of precedence.

LIGHT AND NEAR-INFRARED RADIATION

These TLVs® refer to values for incoherent (non-laser) visible and near-infrared radiation in the wavelength region of 305 to 3000 nm that nearly all workers may be exposed without adverse health effects. The values are based on the best available information from experimental studies. They should be used only as guides in the control of exposures to light and should not be regarded as fine lines between safe and dangerous levels. For purposes of specifying these TLVs®, the optical radiation spectrum is divided into the regions shown in the figure "The Electromagnetic Radiation Spectrum and Related TLVs®" found on page 122.

Recommended Values

The TLVs® for occupational exposure of the eyes to broadband light and near-infrared radiation apply to exposures in any 8-hour workday. Figure 1 is a guide to the application of the TLVs® for visible and near infrared sources.

The first step is to determine if there is a broadband source including the visible light spectrum of sufficient luminance to consider the visible light contributions. If the luminance is greater than 1 candela per square centimeter (cd/cm^2), then the TLVs® in Sections 1 and 2 apply. With a low luminance and no special sources involved, there may not be a significant risk. If the source has a high blue light component such as a blue light-emitting diode (LED), then Section 2 applies. If the source is primarily in the near infrared range because it uses special filters or is in the range by nature (e.g., LED), then Sections 3 and 4 apply. The TLVs® are divided into four potential health effects and spectral regions as follows:

Section 1. *To protect against retinal thermal injury from a visible light source:* Determine the effective spectral radiance of the lamp (L_R) in W/(cm^2 sr) [sr = steradian] by integrating the spectral radiance (L_λ) in W/(cm^2 sr nm) weighted by the thermal hazard function R(λ), using Equation 1 or a light meter with an R(λ) filter. R(λ) is shown in Figure 2 and values are provided in Table 1.

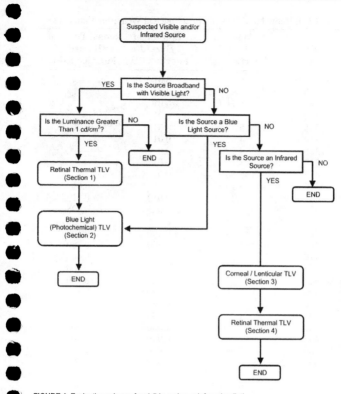

FIGURE 1. Evaluation scheme for visible and near infrared radiation.

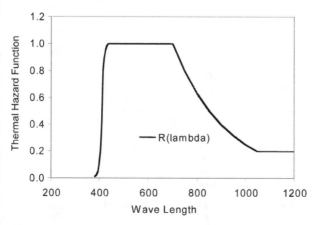

FIGURE 2. Retinal thermal hazard function [R(λ)].

TLV®–PA

TABLE 1. Retinal and UVR Hazard Spectral Weighting Functions

Wavelength (nm)	Aphakic Hazard Function A(λ)	Blue-Light Hazard Function B(λ)	Retinal Thermal Hazard Function R(λ)
305–335	6.00	0.01	—
340	5.88	0.01	—
345	5.71	0.01	—
350	5.46	0.01	—
355	5.22	0.01	—
360	4.62	0.01	—
365	4.29	0.01	—
370	3.75	0.01	—
375	3.56	0.01	—
380	3.19	0.01	0.01
385	2.31	0.0125	0.0125
390	1.88	0.025	0.025
395	1.58	0.050	0.050
400	1.43	0.100	0.100
405	1.30	0.200	0.200
410	1.25	0.400	0.400
415	1.20	0.800	0.800
420	1.15	0.900	0.900
425	1.11	0.950	0.950
430	1.07	0.980	0.980
435	1.03	1.000	1.00
440	1.000	1.000	1.00
445	0.970	0.970	1.00
450	0.940	0.940	1.00
455	0.900	0.900	1.00
460	0.800	0.800	1.00
465	0.700	0.700	1.00
470	0.620	0.620	1.00
475	0.550	0.550	1.00
480	0.450	0.450	1.00
485	0.400	0.400	1.00
490	0.220	0.220	1.00
495	0.160	0.160	1.00
500	0.100	0.100	1.00
505	0.079	0.079	1.00
510	0.063	0.063	1.00
515	0.050	0.050	1.00
520	0.040	0.040	1.00
525	0.032	0.032	1.00
530	0.025	0.025	1.00
535	0.020	0.020	1.00
540	0.016	0.016	1.00
545	0.013	0.013	1.00
550	0.010	0.010	1.00
555	0.008	0.008	1.00
560	0.006	0.006	1.00
565	0.005	0.005	1.00
570	0.004	0.004	1.00
575	0.003	0.003	1.0

TABLE 1 (con't.). Retinal and UVR Hazard Spectral Weighting Functions

Wavelength (nm)	Aphakic Hazard Function A(λ)	Blue-Light Hazard Function B(λ)	Retinal Thermal Hazard Function R(λ)
580	0.002	0.002	1.0
585	0.002	0.002	1.0
590	0.001	0.001	1.0
595	0.001	0.001	1.0
600–700	0.001	0.001	1.0
700–1050	—	—	$10^{[(700-\lambda)/500]}$
1050–1400	—	—	0.2

$$L_R \, [\text{W/(cm2sr)}] = \sum_{380}^{1400} L_\lambda \cdot R(\lambda) \cdot \Delta\lambda \tag{1}$$

Some meters provide a total energy emitted in units of J/(cm² sr) over the sampling period, which is the time integral of L_R over the sampling period. Therefore, an alternative expression of the retinal thermal injury TLV® is a dose limit (called DL_R in this TLV®).

Determine the angular subtense (α) of the source in radians (rad). For circular lamps, α is the lamp diameter divided by the viewing distance. If the lamp is oblong, α is estimated from the mean of the shortest and longest dimension that can be viewed divided by the viewing distance, which is according to Equation 2.

$$\alpha \, [\text{rad}] = \frac{(l + w)}{2r} \tag{2}$$

For instance, at a viewing distance r = 100 cm from a 0.8-cm diameter tubular flash lamp of length l = 5 cm, the viewing angle α is 0.029 rad.

Large sources are those with an angular subtense (α) greater than 0.1 rad. For large sources, Equations 3a through 3c define the TLV® for protection against retinal thermal injury depending on the exposure duration (t) in seconds [s]. These limits also serve as a useful screening step.

For viewing durations (t) from 1 µs (10^{-6} s) through 0.00063 s, an acceptable exposure is present when Equation 3a is true. For pulse durations less than 1 µs, the TLV® is the same as that for 1 µs. Since the retinal thermal hazard TLVs® for pulsed sources assume a 7-mm, dark-adapted pupil, this exposure limit may be modified for daylight conditions.

$$L_R \, [\text{W/(cm}^2 \text{ sr)}] \leq \frac{640}{t^{1/4}} \quad \text{OR}$$

$$DL_R \, [\text{J/(cm}^2 \text{ sr)}] \leq 640 \cdot t^{0.75} \tag{3a}\blacklozenge$$

For viewing durations between 0.63 ms (0.00063 s) and 0.25 s, an acceptable exposure is present when Equation 3b is true.

TLV®–PA

$$L_R \, [W/(cm^2 \, sr)] \leq \frac{16}{t^{0.75}} \quad OR$$

$$DL_R \, [J/(cm^2 \, sr)] \leq 16 \cdot t^{1/4} \tag{3b}\blacklozenge$$

For viewing durations greater than 0.25 s, an acceptable exposure is present when Equation 3c is true. This is a rate-, rather than dose-, limited threshold.

$$L_R \, [W/(cm^2 \, sr)] \leq 45 \tag{3c}\blacklozenge$$

Small sources have an angular subtense (α) less than 0.1 rad. For small sources, the retinal thermal injury risk depends on both the exposure duration (t) and α. The interaction is a maximum value for α (α_{max}) as a function of viewing duration (t [s]).

For viewing durations from 1 μs (10^{-6} s) through 0.00063 s, an acceptable exposure is present when Equation 3a above is true. For pulse durations less than 1 μs, the TLV® is the same as that for 1 μs. Since the retinal thermal hazard TLVs® for pulsed sources assume a 7-mm, dark-adapted pupil, this exposure limit may be modified for daylight conditions.

For viewing durations from 0.00063 to 0.25 s, an acceptable exposure is present when Equation 4a is true.

With $\alpha < \alpha_{max} = 0.2 \cdot t^{0.5}$ rad,

$$L_R \, [W/(cm^2 \, sr)] \leq \frac{3.2}{\alpha \cdot t^{1/4}} \quad OR$$

$$DL_R \, [J/(cm^2 \, sr)] \leq \frac{3.2 \cdot t^{0.75}}{\alpha} \tag{4a}\blacklozenge$$

For viewing durations greater than 0.25 s, an acceptable exposure is present when Equation 4b is true. This is a rate-limited exposure and a dose limit does not apply.

With $\alpha < \alpha_{MAX} = 0.1$ rad,

$$L_R \, [W/(cm^2 \, sr)] \leq \frac{4.5}{\alpha} \tag{4b}\blacklozenge$$

Note: There may be special individual circumstances where the pupil remains dilated (tonic) and exposures extend beyond 0.25 s. Under these conditions, Equation 4c is the limiting exposure.

With $\alpha < \alpha_{MAX} = 0.1$ rad,

$$L_R \, [W/(cm^2 \, sr)] \leq \frac{3.2}{\alpha \cdot t^{1/4}} \tag{4c}\blacklozenge$$

Section 2. *To protect against retinal photochemical injury from chronic blue-light (305 < λ < 700 nm) exposure:* Determine the integrated effective spectral radiance of the light source (L_B) in W/(cm^2 sr) by integrating the spectral radiance (L_λ) in W/(cm^2 sr nm) weighted by the blue-light hazard function B(λ) using Equation 5 or a light meter with a B(λ) filter. B(λ) is shown in Figure 3 and values are provided in Table 1.

$$L_B \, [\text{W/(cm}^2\,\text{sr)}] = \sum_{305}^{700} L_\lambda \cdot B(\lambda) \cdot \Delta\lambda \tag{5}$$

Some meters provide a total energy emitted in units of J/(cm^2 sr) over the sampling period, which is the time integral of L_B over the sampling period. L_B is the total energy divided by the sample period.

For viewing durations (t) less than 10^4 s (167 min or ~ 2.8 h) in a day, an acceptable exposure is present when:

$$L_B \leq \frac{100 \, [\text{J/(cm}^2\,\text{sr)}]}{t\,[\text{s}]} \tag{6a}$$

Alternatively, when L_B exceeds 0.01 W/(cm^2 sr), the maximum acceptable exposure duration t_{max} in seconds is:

$$t_{max} \, [\text{s}] = \frac{100 \, [\text{J/(cm}^2\,\text{sr)}]}{L_B} \tag{6b}$$

For viewing durations greater than 10^4 s (167 min) in a day, an acceptable exposure is present when:

$$L_B \, [\text{W/(cm}^2\,\text{sr)}] \leq 10^{-2} \tag{6c}$$

Note for blue light hazard: The L_B limits are greater than the maximum permissible exposure limits for 440 nm laser radiation (*see* Laser TLV®) because of the need for caution related to narrow-band spectral effects of lasers.

SPECIAL CASE FOR SMALL-SOURCE ANGLES: For a light source subtending an angle less than 0.011 radian, the above limits are relaxed. Determine the spectral irradiance (E_λ) weighted by the blue-light hazard function B(λ):

$$E_B \, [\text{W/cm}^2] = \sum_{305}^{700} E_\lambda \cdot B(\lambda) \cdot \Delta\lambda \tag{7}$$

For durations less than 100 s (1 min, 40 s) in a day, an acceptable exposure is present when:

$$E_B \leq \frac{0.01 \, [\text{J/cm}^2]}{t\,[\text{s}]} \tag{8a}$$

Alternatively, for a source where the blue-light-weighted irradiance E_B exceeds 10^{-4} W/cm^2, the maximum acceptable exposure duration, t_{max}, in seconds is:

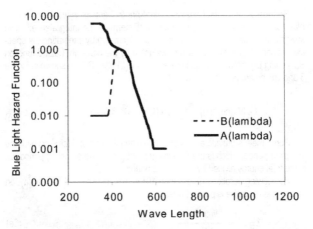

FIGURE 3. Blue light (retinal photochemical) hazard function for normal eyes [B(λ)] and the aphakic hazard function [A(λ)].

$$t_{max} \text{ [s]} = \frac{0.01 \text{ [J/cm}^2]}{E_B} \tag{8b}$$

For viewing durations greater than 10^2 s (1 min, 40 s) in a day, an acceptable exposure is present when:

$$E_B \leq 10^{-4} \text{ [W/cm}^2] \tag{8c}$$

SPECIAL CASE: To protect the worker having a lens removed (cataract surgery) against retinal photochemical injury from chronic exposure: Unless an ultraviolet (UV)-absorbing intra-ocular lens has been surgically inserted into the eye, the Aphakic Hazard Function, A(λ), should be used for L_B and E_B, as shown in Equations 9a and 9b.

$$L_B \text{ [W/(cm}^2 \text{ sr)]} = \sum_{305}^{700} L_\lambda \cdot A(\lambda) \cdot \Delta\lambda \tag{9a}$$

$$E_B \text{ [W/(cm}^2 \text{ sr)]} = \sum_{305}^{700} E_\lambda \cdot A(\lambda) \cdot \Delta\lambda \tag{9b}$$

The value for L_B is used in Equation 6 and the value for E_B is used in Equation 8.

Section 3. *To protect against thermal injury to the cornea and lens from infrared (IR) radiation:* To avoid thermal injury of the cornea and possible delayed effects on the lens of the eye (cataractogenesis), the total infrared irradiance in hot environments is calculated as

$$E_{IR-only} \, [\text{W/cm}^2] = \sum_{770}^{3000} E_\lambda \cdot \Delta\lambda \qquad (10)$$

For exposure durations (t) less than 10^3 sec (17 min), an acceptable exposure is present when:

$$E_{IR-only} \, [\text{W/cm}^2] \leq \frac{1.8}{t^{0.75}} \qquad (11a)$$

For exposure durations greater than 10^3 sec (17 min), an acceptable exposure is present when:

$$E_{IR-only} \, [\text{W/cm}^2] \leq 0.01 \qquad (11b)$$

Section 4. *To protect against retinal thermal injury from near infrared (NIR) radiation:* For a near infrared source associated with an infrared heat lamp or any NIR source where a strong visual stimulus is absent (luminance less than 10^{-2} cd/cm^2), the total effective radiance (L_{NIR}) as viewed by the eye is the spectral radiance (L_λ) weighted by the thermal hazard function, $R(\lambda)$.

$$L_{NIR} \, [\text{W/(cm}^2 \, \text{sr})] = \sum_{770}^{1400} L_\lambda \cdot R(\lambda) \cdot \Delta\lambda \qquad (12)$$

For exposures less than 810 s, an acceptable exposure is present when:

$$L_{NIR} \, [\text{W/(cm}^2 \, \text{sr})] < \frac{3.2}{\alpha \cdot t^{1/4}} \qquad (13a) \blacklozenge$$

This limit is based upon a 7-mm pupil diameter (since the aversion response may not exist due to an absence of light) and a detector field-of-view of 0.011 rad.

For exposures greater than 810 s in a day, an acceptable exposure is present when:

$$L_{NIR} \, [\text{W/(cm}^2 \, \text{sr})] \leq \frac{0.6}{\alpha} \qquad (13b) \blacklozenge$$

♦   Equations 3, 4, and 13 are empirical and are not, strictly speaking, dimensionally correct. To make the equations dimensionally correct, one would have to insert dimensional correction factors in the right-hand numerator in each equation.

* ULTRAVIOLET RADIATION

These TLVs® refer to incoherent ultraviolet (UV) radiation with wavelengths between 180 and 400 nm and represent conditions under which it is believed that nearly all healthy workers may be repeatedly exposed without acute adverse health effects such as erythema and photokeratitis. Some UV sources covered by this TLV® are welding and carbon arcs, gas and vapor discharges, fluorescent, incandescent and germicidal lamps, and solar radiation. Coherent UV radiation from lasers is covered in the TLV® for Lasers.

The TLV® values apply to continuous sources for exposure durations equal to or greater than 0.1 second. The sources may subtend an angle less than 80 degrees at the detector and for those sources that subtend a greater angle need to be measured over an angle of 80 degrees.

The values do not apply to UV radiation exposure of photosensitive individuals or of individuals concomitantly exposed to photo-sensitizing agents (see Note 3). The values for the eye do not apply to aphakes (persons who have had the lens of the eye removed in cataract surgery), for which case, see Light and Near-Infrared Radiation TLVs®.

The TLVs® should be used as guides in the control of exposure to UV sources and should not be regarded as fine lines between safe and dangerous levels.

Threshold Limit Values

The TLVs® for occupational exposure to UV radiation incident upon the skin or the eye follow. The flow chart in Figure 1 provides a map of the UV TLV®.

Broadband UV Sources (180 to 400 nm) — Corneal Hazard

The first step in evaluating broadband UV sources is to determine the effective irradiance (E_{eff}). To determine E_{eff} for a broadband source weighted against the peak of the spectral effectiveness curve (270 nm), Equation 1 should be used.

$$E_{eff} = \sum_{180}^{400} E_\lambda \bullet S(\lambda) \bullet \Delta\lambda \tag{1}$$

where: E_{eff} = effective irradiance relative to a
monochromatic source at 270 nm [W/cm²]

E_λ = spectral irradiance at a center wavelength
[W/(cm² • nm)]

$S(\lambda)$ = relative spectral effectiveness at the center wavelength
[unitless]

$\Delta\lambda$ = bandwidth around the center wavelength [nm]

More practically, E_{eff} can be measured directly with a UV radiometer having a built-in spectral response that mimics the relative spectral effectiveness values in Table 1 and Figure 2.

The daily exposure (t_{exp}) based on E_{eff} is dose limited to 0.003 J/cm². That is,

$$0.003[J/cm^2] \geq E_{eff}[W/cm^2] \bullet t_{exp}[s] \qquad (2)$$

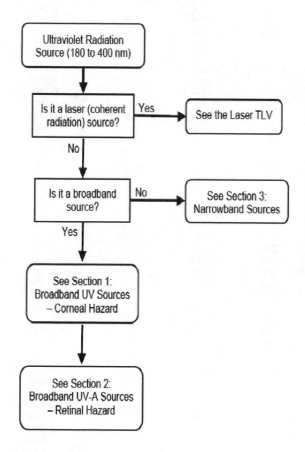

FIGURE 1. Flow chart for UV TLV®.

TABLE 1. Ultraviolet Radiation TLV® and Relative Spectral Effectiveness

Wavelength[A] (nm)	TLV® (J/m^2)[B]	TLV® (mJ/cm^2)[B]	Relative Spectral Effectiveness, $S(\lambda)$
180	2500	250	0.012
190	1600	160	0.019
200	1000	100	0.030
205	590	29	0.051
210	400	40	0.075
215	320	32	0.095
220	250	25	0.120
225	200	20	0.150
230	160	16	0.190
235	130	13	0.240
240	100	10	0.300
245	83	8.3	0.360
250	70	7.0	0.430
254[C]	60	6.0	0.500
255	58	5.8	0.520
260	46	4.6	0.650
265	37	3.7	0.810
270	30	3.0	1.00
275	31	3.1	0.960
280[C]	34	3.4	0.880
285	39	3.9	0.770
290	47	4.7	0.640
295	56	5.6	0.540
297[C]	65	6.5	0.460
300	100	10	0.300
303[C]	250	25	0.120
305	500	50	0.060
308	1200	120	0.026
310	2000	200	0.015
313[C]	5000	500	0.006
315	1.0×10^4	1.0×10^3	0.003
316	1.3×10^4	1.3×10^3	0.0024
317	1.5×10^4	1.5×10^3	0.0020
318	1.9×10^4	1.9×10^3	0.0016
319	2.5×10^4	2.5×10^3	0.0012
320	2.9×10^4	2.9×10^3	0.0010
322	4.5×10^4	4.5×10^3	0.00067
323	5.6×10^4	5.6×10^3	0.00054
325	6.0×10^4	6.0×10^3	0.00050
328	6.8×10^4	6.8×10^3	0.00044
330	7.3×10^4	7.3×10^3	0.00041
333	8.1×10^4	8.1×10^3	0.00037
335	8.8×10^4	8.8×10^3	0.00034
340	1.1×10^5	1.1×10^4	0.00028
345	1.3×10^5	1.3×10^4	0.00024
350	1.5×10^5	1.5×10^4	0.00020

TABLE 1 (con't.). Ultraviolet Radiation TLV® and Relative Spectral Effectiveness

Wavelength[A] (nm)	TLV® (J/m^2)[B]	TLV® (mJ/cm^2)[B]	Relative Spectral Effectiveness, $S(\lambda)$
355	1.9×10^5	1.9×10^4	0.00016
360	2.3×10^5	2.3×10^4	0.00013
365[C]	2.7×10^5	2.7×10^4	0.00011
370	3.2×10^5	3.2×10^4	0.000093
375	3.9×10^5	3.9×10^4	0.000077
380	4.7×10^5	4.7×10^4	0.000064
385	5.7×10^5	5.7×10^4	0.000053
390	6.8×10^5	6.8×10^4	0.000044
395	8.3×10^5	8.3×10^4	0.000036
400	1.0×10^6	1.0×10^5	0.000030

[A] Wavelengths chosen are representative; other values should be interpolated at intermediate wavelengths.

[B] 1 mJ/cm² = 10 J/m²

[C] Emission lines of a mercury discharge spectrum.

Table 2 gives TLV® values for the effective irradiance for different daily exposure durations. In general, the maximum exposure time (t_{max}) [s] for a broadband UV source can be determined from Equation 3.

$$t_{max}[s] = \frac{0.003\,[J/cm^2]}{E_{eff}\,[W/cm^2]} \tag{3}$$

TLV®-PA

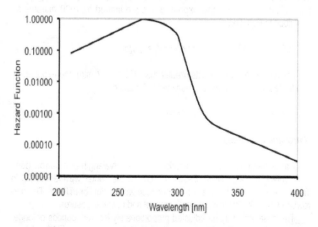

FIGURE 2. Hazard function (relative spectral effectiveness, $S(\lambda)$ for UV.

TABLE 2. Exposure Durations for Given Actinic UV Radiation Effective Irradiances

Duration of Exposure Per Day	Effective Irradiance, E_{eff} (mW/cm²)
8 hours	0.0001
4 hours	0.0002
2 hours	0.0004
1 hour	0.0008
30 minutes	0.0017
15 minutes	0.0033
10 minutes	0.005
5 minutes	0.01
1 minute	0.05
30 seconds	0.1
10 seconds	0.3
1 second	3
0.5 second	6
0.1 second	30

Broadband UV-A Sources (315 to 400 nm) — Lens and Retinal Hazard

The irradiance, E_{UV-A} [mW/cm²], can be measured with an unfiltered meter that is sensitive to UV-A radiation. For daily exposure periods (t_{exp}) less than 1000 s (17 min), the exposure is dose limited to 1000 mJ/cm² as described in Equation 4.

$$1000[mJ/cm^2] \geq E_{UV-A}[mW/cm^2] \bullet t_{exp}[s] \tag{4}$$

For daily exposure periods greater than 1000 s (17 min), the exposure is rate limited to 1.0 mW/cm² as described in Equation 5.

$$1.0[mW/cm^2] \geq E_{UV-A}[mW/cm^2] \tag{5}$$

Narrowband Sources

Narrowband sources are comprised of one wavelength or a narrow band of wavelengths (e.g., within 5–10 nm). Locate the center wavelength (λ) in Table 1, and find the TLV_λ as an 8-hour dose limit in J/m² or mJ/cm². The narrowband TLV® is protective for both corneal and retinal exposures.

The dose limit may be adjusted proportionally for work periods of longer or shorter duration. The TLV® dose limit of a daily exposure period (t_{exp}) for a narrowband source can be expressed as Equation 6 using the Spectral Sensitivity (S_λ) from Table 1 and unfiltered irradiance (E_λ) [W/m² or mW/cm²].

$$30[J/m^2] \geq E_\lambda \, [W/m^2] \bullet S(\lambda) \bullet t_{exp}[s] \qquad (6a)$$

$$3.0[mJ/cm^2] \geq E_\lambda \, [mW/cm^2] \bullet S(\lambda) \bullet t_{exp}[s] \qquad (6b)$$

The maximum exposure time (t_{max}) [s] for a narrowband source can be determined from Equation 7 using the TLV_λ and the unfiltered irradiance (E_λ) [W/m² or mW/cm²]. (Note: The energy and surface area units must match.)

$$t_{max}[s] = \frac{TLV_\lambda}{E_\lambda} \qquad (7)$$

Notes:

1. The probability of developing skin cancer depends on a variety of factors such as skin pigmentation, a history of blistering sunburns, and the accumulated UV dose. It also depends on genetic susceptibility and factors such as skin and eye color. Individuals who have a familial history of melanoma, or numerous nevi over their body, for example, may be at higher risk of developing malignant melanoma. The risks for developing melanoma and non-melanoma cancers may differ from each other and depend on the UV exposure history.

2. Outdoor workers in latitudes within 40 degrees of the equator can be exposed outdoors to levels above the TLVs® in as little as five minutes around noontime during the summer.

3. Exposure to ultraviolet radiation concurrently with topical or systemic exposure to a variety of chemicals, including some prescription drugs, can result in skin erythema at sub-TLV® exposures. Hypersensitivity should be suspected if workers present skin reactions when exposed to sub-TLV® doses or when exposed to levels (generally UV-A) that did not cause a noticeable erythema in the same individual in the past. Among the hundreds of agents that can cause hypersensitivity to UV radiation are certain plants and chemicals such as some antibiotics (e.g., tetracycline and sulphathiazole), some antidepressants (e.g., imipramine and sinequan), as well as some diuretics, cosmetics, antipsychotic drugs, coal tar distillates, some dyes, or lime oil.

4. Ozone is produced in air by sources emitting UV radiation at wavelengths below 250 nm. Refer to the latest version of the Chemical Substances TLV® for ozone.

‡ LASERS

These TLVs® are for exposure to laser radiation under conditions to which it is believed nearly all workers may be repeatedly exposed without adverse health effects. The TLVs® should be used as guides in the control of exposures and should not be regarded as fine lines between safe and dangerous levels. They are based on the best available information from experimental studies. In practice, hazards to the eye and skin can be controlled by application of control measures appropriate to the classification of the laser.

Classification of Lasers

Most lasers have a label affixed to them by the manufacturer that describes their hazard class. Normally, it is not necessary to determine laser irradiances or radiant exposures for comparison with the TLVs®. The potential for hazardous exposures can be minimized by the application of control measures that are appropriate to the hazard class of the laser. Control measures are applicable to all classes of lasers except for Class 1. Such measures, and other laser safety information, may be found in the ACGIH® publication, *A Guide for Control of Laser Hazards*, and the ANSI Z136 series published by the Laser Institute of America.

Limiting Apertures

For comparison with the TLVs® in this section, laser beam irradiance or radiant exposure is averaged over the limiting aperture appropriate to the spectral region and exposure duration. If the laser beam diameter is less than that of the limiting aperture, the effective laser beam irradiance or radiant exposure may be calculated by dividing the laser beam power or energy by the area of the limiting aperture. Limiting apertures are listed in Table 1.

TABLE 1. Limiting Apertures Applicable to Laser TLVs®

Spectral Region	Duration	Eye	Skin
180 nm–400 nm	1 ns to 0.25 s	1 mm	3.5 mm
180 nm–400 nm	0.25 s to 30 ks	3.5 mm	3.5 mm
400 nm–1400 nm	10^{-4} ns to 0.25 s	7 mm	3.5 mm
400 nm–1400 nm	0.25 s to 30 ks	7 mm	3.5 mm
1400 nm–0.1 mm	10^{-5} ns to 0.25 s	1 mm	3.5 mm
1400 nm–0.1 mm	0.25 s to 30 ks	3.5 mm	3.5 mm
0.1 mm–1.0 mm	10^{-5} ns to 30 ks	11 mm	11 mm

TLV®-PA

Repetitively Pulsed Exposures

Scanned, continuous-wave (CW) lasers or repetitively pulsed lasers can both produce repetitively pulsed exposure conditions. The TLV® for intrabeam viewing, which is applicable to wavelengths between 400 and 1400 nm and a single-pulse exposure (of pulse duration t), is modified in this instance by a correction factor determined by the number of pulses in the exposure. First, calculate the number of pulses (n) in an expected exposure situation; this is the pulse repetition frequency (PRF in Hz) multiplied by the duration of exposure. Normally, realistic exposures may range from 0.25 second (s) for a bright, visible source to 10 s for an infrared source. The corrected TLV® on a per-pulse basis is:

$$TLV = (n^{-1/4}) \text{ (TLV for Single-pulse)} \tag{1}$$

This approach applies only to thermal-injury conditions, i.e., all exposures at wavelengths > 700 nm and for many exposures at shorter wavelengths. For wavelengths ≤ 700 nm, the corrected TLV® from Equation 1 applies if the average irradiance does not exceed the TLV® for continuous exposure. The average irradiance (i.e., the total accumulated exposure for nt s) shall not exceed the radiant exposure given in Table 2 for exposure durations of 10 s to T_1.

It is recommended that the user of the TLVs® for laser radiation consult *A Guide for Control of Laser Hazards*, 4th Edition, 1990, published by ACGIH®, for additional information.

TLV®-PA

TLV®–PA

TABLE 2. TLVs® for Direct Ocular Exposures (Intrabeam "Point Source" Viewing) from a Laser Beam

Spectral Region	Wavelength	Exposure, (t) Seconds	TLV®	
UVC	180 nm to 280 nm[α]	10^{-9} to 3×10^{-4}	3 mJ/cm²	
UVB	280 nm to 302 nm	"	3	
	303 nm	"	4	
	304 nm	"	6	
	305 nm	"	10	
	306 nm	"	16	not to exceed (NTE) 0.56 $t^{1/4}$ J/cm² for t ≤ 10 s
	307 nm	"	25	
	308 nm	"	40	
	309 nm	"	63	
	310 nm	"	100	
	311 nm	"	160	
	312 nm	"	250	
	313 nm	"	400	
	314 nm	"	630	
UVA	315 nm to 400 nm	10^{-9} to 10	0.56$t^{1/4}$ J/cm²	
	"	10 to 10^3	1.0 J/cm²	
	"	10^3 to 3×10^4	1.0 mW/cm²	

TABLE 2 (con't.). TLVs® for Direct Ocular Exposures (Intrabeam "Point Source" Viewing) from a Laser Beam

Spectral Region	Wavelength	Exposure, (t) Seconds	TLV®
Light	400 to 700 nm	10^{-13} to 10^{-11}	15 nJ/cm^2
	400 to 700 nm	10^{-11} to 10^{-9}	$2.7\ t^{0.75}$ J/cm^2
	400 to 700 nm	10^{-9} to 18×10^{-6}	5.0×10^{-1} µJ/cm^2
	400 to 700 nm	18×10^{-6} to 10	$1.8\ t^{0.75}$ mJ/cm^2
	400 to 450 nm	10 to 100	10 mJ/cm^2
	450 to 500 nm	10 to 100	1 mW/cm^2
	450 to 500 nm	10 to T_1	$10\ C_B$ mJ/cm^2
	400 to 500 nm	T_1 to 100	$0.1\ C_B$ mW/cm^2
	500 to 700 nm	100 to 3×10^4	1.0 mW/cm^2
IRA	700 to 1050 nm	10^{-13} to 10^{-11}	$15\ C_A$ nJ/cm^2
	700 to 1050 nm	10^{-11} to 10^{-9}	$2.7\ C_A\ t^{0.75}$ J/cm^2
	700 to 1050 nm	10^{-9} to 18×10^{-6}	$0.5\ C_A$ µJ/cm^2
	700 to 1050 nm	18×10^{-6} to 10	$1.8\ C_A\ t^{0.75}$ mJ/cm^2
	700 to 1050 nm	10 to 3×10^4	C_A mW/cm^2
	1050 to 1400 nm	10^{-13} to 10^{-11}	$1.5\ C_C \times 10^{-1}$ µJ/cm^2
	1050 to 1400 nm	10^{-11} to 10^{-9}	$27.0\ C_C\ t^{0.75}$ J/cm^2
	1050 to 1400 nm	10^{-9} to 50×10^{-6}	$5.0\ C_C$ µJ/cm^2
	1050 to 1400 nm	50×10^{-6} to 10	$9.0\ C_C\ t^{0.75}$ mJ/cm^2 NTE 1.0 J/cm^2
	1050 to 1400 nm	10 to 3×10^4	$5.0\ C_C$ mW/cm^2 NTE 100 mW/cm^2

TLV®-PA

TLV®–PA

TABLE 2 (con't). TLVs® for Direct Ocular Exposures (Intrabeam "Point Source" Viewing) from a Laser Beam

Spectral Region	Wavelength	Exposure, (t) Seconds	TLV®
IRB	1.401 to 1.5 μm	10^{-14} to 10^{-3}	0.1 J/cm²
& IRC	1.401 to 1.5 μm	10^{-3} to 10	0.56 $t^{1/4}$ J/cm²
	1.501 to 1.8 μm	10^{-14} to 10	1.0 J/cm²
	1.801 to 2.6 μm	10^{-14} to 10^{-3}	0.1 J/cm²
	1.801 to 2.6 μm	10^{-3} to 10	0.56 $t^{1/4}$ J/cm²
	2.601 to 10^3 μm	10^{-14} to 10^{-7}	10 mJ/cm²
	2.601 to 10^3 μm	10^{-7} to 10	0.56 $t^{1/4}$ J/cm²
	1.400 to 10^3 μm	10 to 3×10^4	100 mW/cm²

*Ozone (O_3) is produced in air by sources emitting ultraviolet (UV) radiation at wavelengths below 250 nm. Refer to Chemical Substances TLV® for ozone.

Notes for Table 2

C_A = Fig. 2; C_B = 1 for λ = 400 to $\leq$ 450 nm; C_B = $10^{[0.02(\lambda - 450)]}$ for λ = 450 to 600 nm; C_C = 1.0 from 700 to 1150 nm; C_C = $10^{[0.018(\lambda - 1150)]}$ for wavelengths greater than 1150 nm and less than 1200 nm; C_C = 8.0 from 1200 to 1300 nm; C_C = 8 + $10^{[0.04(\lambda - 1300)]}$ from 1300 nm to 1400 nm.

T_1 = 10 s for λ = 400 to 450 nm; T_1 = $10 \times 10^{[0.02(\lambda - 550)]}$ for λ = 450 to 500 nm; and T_1 = 10 s for λ = 500 to 700

For intermediate or large sources (e.g., laser diode arrays) at wavelengths between 400 nm and 1400 nm, the intrabeam viewing TLVs® can be increased by correction factor C_E (use Table 3) provided that the angular subtense α of the source (measured at the viewer's eye) is greater than α_{min}. C_E depends on α as follows:

Angular Subtense	Source Size Designation	Correction Factor C_E
$\alpha \leq \alpha_{min}$	Small	$C_E = 1$
$\alpha_{min} < \alpha \leq \alpha_{max}$	Intermediate	$C_E = \alpha/\alpha_{min}$
$\alpha > \alpha_{max}$	Large	$C_E = \alpha_{max}/\alpha_{min} = 3.33$ for $t \leq 0.625$ ms; = $133.33 \ t^{1/2}$ for 0.625 ms < t < 0.25 s = 66.7 for t $\geq$ 0.25 s

The angle referred to as α_{max} corresponds to the point where the TLVs® may be expressed as a constant radiance and the last equation can be rewritten in terms of radiance L.

L_{TLV} = $(3.81 \times 10^5) \times (TLV_{pt \ source}) \ J/(cm^2 \ sr)$ for t < 0.625 μs for 400 < λ < 700 nm

L_{TLV} = 7.6 $t^{1/4} \ J/(cm^2 \ sr)$ for 0.625 ms < t < 0.25 s for 400 < λ < 700 nm

L_{TLV} = 4.8 $W/(cm^2 \ sr)$ for t >100 s for 400 < λ < 700 nm

Figure 5 illustrates these TLVs® for large sources expressed in terms of radiance.

The measurement aperture should be placed at a distance of 100 mm or greater from the source. For large area irradiation, the reduced TLV® for skin exposure applies as noted in the footnote to "IRB & C," Table 4.

TLV®–PA

TLV®-PA

TABLE 3. TLVs® for Extended Source Laser Viewing Conditions

Spectral Region	Wavelength	Exposure, (t) seconds	TLV®
Light	400 to 700 nm	10^{-13} to 10^{-11}	$1.5\,C_E\,10^{-8}$ J/cm²
	400 to 700 nm	10^{-11} to 10^{-9}	$2.7\,C_E\,t^{0.75}$ J/cm²
	400 to 700 nm	10^{-9} to 18×10^{-6}	$5.0\,C_E \times 10^{-7}$ J/cm²
	400 to 700 nm	18×10^{-6} to 0.7	$1.8\,C_E\,t^{0.75} \times 10^{-3}$ J/cm²

Dual Limits for 400 to 600 nm visible laser exposure for t > 0.7 s

Photochemical

For $\alpha \leq 11$ mrad, the MPE is expressed as irradiance and radiant exposure*

	Wavelength	Exposure, (t) seconds	TLV®
	400 to 600 nm	0.7 to 100	$C_B \times 10^{-2}$ J/cm²
	400 to 600 nm	100 to 3 ×10⁴	$C_B \times 10^{-4}$ W/cm²

For $\alpha > 11$ mrad, the MPE is expressed as radiance and integrated radiance*

	Wavelength	Exposure, (t) seconds	TLV®
	400 to 600 nm	0.7 to 1 ×10⁴	$100\,C_B$ J/(cm² sr)
	400 to 600 nm	1 ×10⁴ to 3 ×10⁴	$C_B \times 10^{-2}$ W/(cm² sr)

and

Thermal

	Wavelength	Exposure, (t) seconds	TLV®
	400 to 700 nm	0.7 to T_2	$1.8\,C_E\,t^{0.75} \times 10^{-3}$ J/cm²
	400 to 700 nm	T_2 to 3 ×10⁴	$1.8\,C_E\,T_2^{-0.25} \times 10^{-3}$ W/cm²

TABLE 3 (con't). TLVs® for Extended Source Laser Viewing Conditions

Spectral Region	Wavelength	Exposure, (t) seconds	TLV®
IRA	700 to 1050 nm	10^{-13} to 10^{-11}	$1.5\,C_A\,C_E \times 10^{-8}$ J/cm²
	700 to 1050 nm	10^{-11} to 10^{-9}	$2.7\,C_A\,C_E\,t^{0.75}$ J/cm²
	700 to 1050 nm	10^{-9} to 18×10^{-6}	$5.0\,C_A\,C_E \times 10^{-7}$ J/cm²
	700 to 1050 nm	18×10^{-6} to T_2	$1.8\,C_A\,C_E\,t^{0.75} \times 10^{-3}$ J/cm²
	700 to 1050 nm	T_2 to 3×10^4	$1.8\,C_A\,C_E\,T_2^{-0.25} \times 10^{-3}$ W/cm²
	1050 to 1400 nm	10^{-13} to 10^{-11}	$1.5\,C_C\,C_E \times 10^{-7}$ J/cm²
	1050 to 1400 nm	10^{-11} to 10^{-9}	$27.0\,C_C\,C_E\,t^{0.75}$ J/cm²
	1050 to 1400 nm	10^{-9} to 50×10^{-6}	$5.0\,C_C\,C_E \times 10^{-6}$ J/cm²
	1050 to 1400 nm	50×10^{-6} to T_2	$9.0\,C_C\,C_E\,t^{0.75} \times 10^{-3}$ J/cm² NTE 1.0 J/cm²
	1050 to 1400 nm	T_2 to 3×10^4	$9.0\,C_C\,C_E\,T_2^{-0.25} \times 10^{-3}$ W/cm² NTE 1.0 J/cm²

* For sources subtending an angle greater than 11 mrad, the limit may also be expressed as an integrated radiance $L_p =$ $100\,C_B$ J/(cm² sr) for $0.7\,s \le t < 10^4$ s and $L_e = C_B \times 10^{-2}$ W/(cm² sr) for $t \ge 10^4$ s as measured through a limiting cone angle γ.

TABLE 3 (con't.). TLVs® for Extended Source Laser Viewing Conditions

These correspond to values of J/cm^2 for $10 \text{ s} \leq t < 100$ s and W/cm^2 for $t \geq 100$ s as measured through a limiting cone angle γ.

$\gamma = 11$ mrad for $0.7 \text{ s} \leq t < 100$ s
$\gamma = 1.1 \times t^{0.5}$ mrad for $100 \leq t < 10^4$ s
$\gamma = 110$ mrad for $10^4 \text{ s} \leq t < 3 \times 10^4$ s

$\tau_2 = 10 \times 10^{(\alpha - 1.5)/98.5}$ for α expressed in mrad for $\lambda = 400$ to 1400 nm.

For exposure duration "t", the angle α_{max} is defined as:

$\alpha_{max} = 5$ mrad for $t \leq$ to 0.625 ms
$\alpha_{max} = 200 \, t^{0.5}$ mrad for 0.625 ms $< t < 0.25$ s, and
$\alpha_{max} = 100$ mrad for $t \geq 0.25$ s
$\alpha_{min} = 1.5$ mrad

TABLE 4. TLVs® for Skin Exposure from a Laser Beam

Spectral Region	Wavelength	Exposure, (t) Seconds	TLV®
UVA[A]	180 nm to 400 nm	10^{-9} to 10^4	Same as Table 2
Light & IRA	400 nm to 1400 nm	10^{-9} to 10^{-7}	$2\,C_A \times 10^{-2}$ J/cm²
	" "	10^{-7} to 10	$1.1\,C_A\,\sqrt[4]{t}$ J/cm²
	" "	10 to 3×10^4	$0.2\,C_A$ W/cm²
IRB & C[B]	1.401 to 10^3 μm	10^{-14} to 3×10^4	Same as Table 2

[A] Ozone (O_3) is produced in air by sources emitting ultraviolet (UV) radiation at wavelengths below 250 nm. Refer to Chemical Substances TLV® for ozone.

$$C_A = 1.0 \text{ for } \lambda = 400 - 700 \text{ nm; see Figure 2 for } \lambda = 700 \text{ to } 1400 \text{ nm}$$

[B] At wavelengths greater than 1400 nm, for beam cross-sectional areas exceeding 100 cm², the TLV® for exposure durations exceeding 10 seconds is:

$$TLV = (10,000/A_s) \text{ mW/cm}^2$$

where A_s is the irradiated skin area for 100 to 1000 cm², and the TLV® is 10 mW/cm² for irradiated skin areas exceeding 1000 cm² and is 100 mW/cm² for irradiated skin areas less than 100 cm².

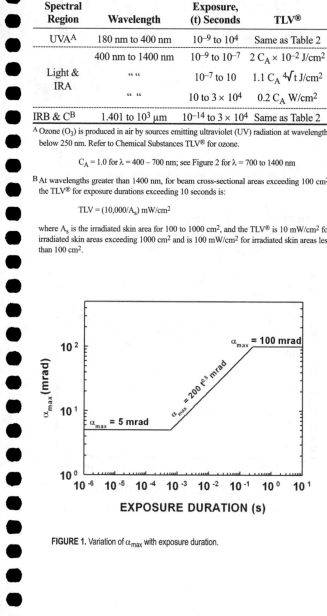

FIGURE 1. Variation of α_{max} with exposure duration.

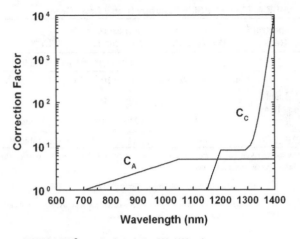

FIGURE 2. TLV® correction factor for λ = 700–1400 nm*
(*For λ = 700–1049 nm, $C_A = 10^{[0.002(\lambda - 700)]}$; for λ = 1050–1400 nm, $C_A = 5$. For λ ≤ 1150, $C_C = 1$; for λ = 1150–1200, $C_C = 10^{0.018(\lambda - 1150)}$; for λ = 1200–1300, $C_C = 8$; and for λ = 1200–1300, $C_C = 8 + 10^{0.04(\lambda - 1300)}$).

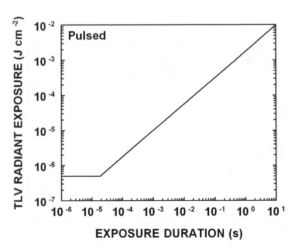

FIGURE 3a. TLV® for intrabeam (direct) viewing of laser beam (400–700 nm).

TLV®–PA

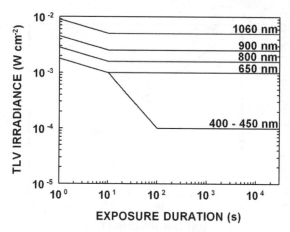

FIGURE 3b. TLV® for intrabeam (direct) viewing of CW laser beam (400–1400 nm).

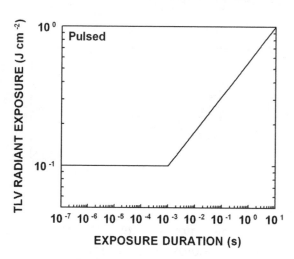

FIGURE 4a. TLV® for laser exposure of skin and eyes for far-infrared radiation (wavelengths greater than 1.4 µm).

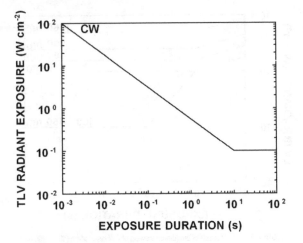

FIGURE 4b. TLV® for CW laser exposure of skin and eyes for far-infrared radiation (wavelengths greater than 1.4 µm).

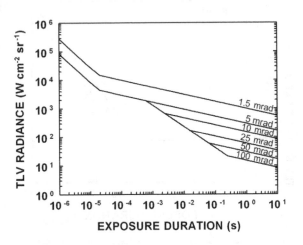

FIGURE 5. TLVs® in terms of radiance for exposures to extended-source lasers in the wavelength range of 400 to 700 nm.

NOTICE OF INTENDED CHANGE —
† LASERS

The reason for this NIC is to add notes to Tables 2 and 3 "NTE" dual limits; to revise the TLVs® for pulse durations less than 50 ns and TLVs® between 1.4 and 1.5 μm; and to revise C_c.

These TLVs® are for exposure to laser radiation under conditions to which it is believed nearly all workers may be repeatedly exposed without adverse health effects. The TLVs® should be used as guides in the control of exposures and should not be regarded as fine lines between safe and dangerous levels. They are based on the best available information from experimental studies. In practice, hazards to the eye and skin can be controlled by application of control measures appropriate to the classification of the laser.

Classification of Lasers

Most lasers have a label affixed to them by the manufacturer that describes their hazard class. Normally, it is not necessary to determine laser irradiances or radiant exposures for comparison with the TLVs®. The potential for hazardous exposures can be minimized by the application of control measures that are appropriate to the hazard class of the laser. Control measures are applicable to all classes of lasers except for Class 1. Such measures, and other laser safety information, may be found in the ACGIH® publication, *A Guide for Control of Laser Hazards*, and the ANSI Z136 series published by the Laser Institute of America.

Limiting Apertures

For comparison with the TLVs® in this section, laser beam irradiance or radiant exposure is averaged over the limiting aperture appropriate to the spectral region and exposure duration. If the laser beam diameter is less than that of the limiting aperture, the effective laser beam irradiance or radiant exposure may be calculated by dividing the laser beam power or energy by the area of the limiting aperture. Limiting apertures are listed in Table 1.

Source Size and Correction Factor C_E

The following considerations apply only at wavelengths in the retinal hazard region, 400–1400 nanometers (nm). Normally, a laser is a small source, which approximates a "point" source and subtends an angle less than α_{min}, which is 1.5 mrad for all values of t. However, any source that subtends an angle, α, greater than α_{min}, and is measured from the viewer's eye, is treated as an "intermediate source" ($\alpha_{min} < \alpha \leq \alpha_{max}$) or a "large, extended source" ($\alpha > \alpha_{max}$). For exposure duration "t", the angle α_{max} is defined as:

α_{max} = 5 mrad for t ≤ 0.625 ms
α_{max} = 200 • $t^{0.5}$ mrad for 0.625 ms < t < 0.5 s
α_{max} = 100 mrad for t ≥ 0.25 s, and
α_{min} = 1.5 mrad

Figure 1 illustrates the time dependence of a_{max}. If the source is oblong, alpha is determined from the arithmetic average of the longest and shortest viewable dimensions.

TLV®-PA

For intermediate and large sources, the TLVs® in Table 2 are modified by a correction factor C_E, as detailed in the Notes for Table 2.

Correction Factors A, B, C (C_A, C_B, C_C)

The TLVs® for ocular exposures in Table 2 are to be used as given for all wavelength ranges. The TLVs® for wavelengths between 700 and 1049 nm are to be increased by the factor C_A (to account for reduced absorption of melanin) as given in Figure 2. For certain exposure times at wavelengths between 400 and 600 nm, a correction factor C_B (to account for reduced photochemical sensitivity for retinal injury) is applied. The correction factor C_C is applied from 1150 to 1400 nm to account for pre-retinal absorption of the ocular media.

The TLVs® for skin exposure are given in Table 4. The TLVs® are to be increased by a factor C_A, as shown in Figure 2, for wavelengths between 700 nm and 1400 nm. To aid in the determination for exposure durations requiring calculations of fractional powers, Figures 3a, 3b, 4a, and 4b may be used.

Repetitively Pulsed Exposures

Scanned, continuous-wave (CW) lasers or repetitively pulsed lasers can both produce repetitively pulsed exposure conditions. The TLV® for intrabeam viewing, which is applicable to wavelengths between 400 and 1400 nm and a single-pulse exposure (of pulse duration t), is modified in this instance by a correction factor determined by the number of pulses in the exposure. First, calculate the number of pulses (n) in an expected exposure situation; this is the pulse repetition frequency (PRF in Hz) multiplied by the duration of exposure. Normally, realistic exposures may range from 0.25 second (s) for a bright, visible source to 10 s for an infrared source. The corrected TLV® on a per-pulse basis is:

$$TLV = (n^{-1/4})(TLV \text{ for Single-pulse}) \qquad (1)$$

This approach applies only to thermal-injury conditions, i.e., all exposures at wavelengths > 700 nm and for many exposures at shorter wavelengths. For wavelengths ≤ 700 nm, the corrected TLV® from Equation 1 applies if the average irradiance does not exceed the TLV® for continuous exposure. The average irradiance (i.e., the total accumulated exposure for nt s) shall not exceed the radiant exposure given in Table 2 for exposure durations of 10 s to T1. It is recommended that the user of the TLVs® for laser radiation consult *A Guide for Control of Laser Hazards*, 4th Edition, 1990, published by ACGIH®, for additional information.

TABLE 1. Limiting Apertures Applicable to Laser TLVs®

Spectral Region	Duration	Eye	Skin
180 nm–400 nm	1 ns to 0.25 s	1 mm	3.5 mm
180 nm–400 nm	0.25 s to 30 ks	3.5 mm	3.5 mm
400 nm–1400 nm	10^{-4} ns to 0.25 s	7 mm	3.5 mm
400 nm–1400 nm	0.25 s to 30 ks	7 mm	3.5 mm
1400 nm–0.1 mm	10^{-5} ns to 0.25 s	1 mm	3.5 mm
1400 nm–0.1 mm	0.25 s to 30 ks	3.5 mm	3.5 mm
0.1 mm–1.0 mm	10^{-5} ns to 30 ks	11 mm	11 mm

TABLE 2. TLVs® for Direct Ocular Exposures (Intrabeam "Point Source" Viewing) from a Laser Beam

Spectral Region	Wavelength	Exposure, (t) Seconds	TLV®
UVC	180 nm to 280 nm*	10^{-9} to 3×10^4	3 mJ/cm²
UVB	280 nm to 302 nm	"	3 mJ/cm²
	303 nm	"	4 mJ/cm²
	304 nm	"	6 mJ/cm²
	305 nm	"	10 mJ/cm²
	306 nm	"	16 mJ/cm²
	307 nm	"	25 mJ/cm²
	308 nm	"	40 mJ/cm²
	309 nm	"	63 mJ/cm²
	310 nm	"	100 mJ/cm²
	311 nm	"	160 mJ/cm²
	312 nm	"	250 mJ/cm²
	313 nm	"	400 mJ/cm²
	314 nm	"	630 mJ/cm²
UVA	315 nm to 400 nm	10^{-9} to 10	$0.56\ t^{1/4}$ J/cm²
	315 nm to 400 nm	10 to 10^3	1.0 J/cm²
	315 nm to 400 nm	10^3 to 3×10^4	1.0 mW/cm²

Not to exceed (NTE) $0.56\ t^{1/4}$ J/cm² for t ≤ 10 s

TLV®–PA

TABLE 2 (con't). TLVs® for Direct Ocular Exposures (Intrabeam "Point Source" Viewing) from a Laser Beam

Spectral Region	Wavelength	Exposure, (t) Seconds	TLV®
Light	400 to 700 nm	10^{-13} to 23×10^{-9}	5×10^{-8} J/cm²
	400 to 700 nm	23×10^{-9} to 50×10^{-9}	4×10^{15} t^3 J/cm²
	400 to 700 nm	50×10^{-9} to 18×10^{-6}	5.0×10^{-7} J/cm²
	400 to 700 nm	18×10^{-6} to 10	1.8 $t^{0.75}$ mJ/cm²
	400 to 450 nm	10 to 100	10 mJ/cm²
	450 to 500 nm	10 to T_1	1 mW/cm²
	450 to 500 nm	T_1 to 100	10 C_B mJ/cm²
	400 to 500 nm	100 to 3×10^4	0.1 C_B mW/cm²
	500 to 700 nm	10 to 3×10^4	1.0 mW/cm²
IRA	700 to 1050 nm	10^{-13} to 5×10^{-13}	5×10^{-8} J/cm²
	700 to 1050 nm	5×10^{-13} to 23×10^{-9}	5 $C_A \times 10^{-8}$ J/cm²
	700 to 1050 nm	23×10^{-9} to 50×10^{-9}	4 $C_A \times 10^{15}$ t^3 J/cm²
	700 to 1050 nm	50×10^{-9} to 18×10^{-6}	5 $C_A \times 10^{-7}$ J/cm²
	700 to 1050 nm	18×10^{-6} to 10	1.8 C_A $t^{0.75} \times 10^{-3}$ J/cm²
	700 to 1050 nm	10 to 3×10^4	$C_A \times 10^{-3}$ W/cm²
	1050 to 1400 nm	10^{-13} to 5×10^{-13}	$C_C \times 10^{-7}$ J/cm²
	1050 to 1400 nm	5×10^{-13} to 25×10^{-9}	5.0 $C_C \times 10^{-7}$ J/cm²
	1050 to 1400 nm	25×10^{-9} to 50×10^{-9}	40 $C_C \times 10^{15}$ t^3 J/cm²
	1050 to 1400 nm	50×10^{-9} to 50×10^{-6}	5.0 $C_C \times 10^{-6}$ J/cm²
	1050 to 1400 nm	50×10^{-6} to 10	9.0 C_C $t^{0.75} \times 10^{-3}$ J/cm²
	1050 to 1400 nm	10 to 3×10^4	5.0 $C_C \times 10^{-3}$ W/cm²

TABLE 2 (con't.). TLVs® for Direct Ocular Exposures (Intrabeam "Point Source" Viewing) from a Laser Beam

Spectral Region	Wavelength	Exposure, (t) Seconds	TLV®
IRB	1.401 to 1.5 μm	10^{-14} to 10^{-3}	0.3 J/cm^2
& IRC	1.401 to 1.5 μm	10^{-3} to 4.0	$0.56\,t^{0.25} + 0.2$ J/cm^2
	1.401 to 1.5 μm	4.0 to 10	1.0 J/cm^2
	1.501 to 1.8 μm	10^{-14} to 10	1.0 J/cm^2
	1.801 to 2.6 μm	10^{-14} to 10^{-3}	0.1 J/cm^2
	1.801 to 2.6 μm	10^{-3} to 10	$0.56\,t^{1/4}$ J/cm^2
	2.601 to 10^3 μm	10^{-14} to 10^{-7}	10 mJ/cm^2
	2.601 to 10^3 μm	10^{-7} to 10	$0.56\,t^{1/4}$ J/cm^2
	1.400 to 10^3 μm	10 to 3×10^4	100 mW/cm^2

*Ozone (O_3) is produced in air by sources emitting ultraviolet (UV) radiation at wavelengths below 250 nm. Refer to Chemical Substances TLV® for ozone.

TLV®-PA

Notes for Table 2

C_A = Fig. 2; C_B = 1 for λ = 400 to $\leq$ 450 nm; C_B = $10^{0.02(\lambda - 450)}$ for λ = 450 to 600 nm; C_C = 1.0 for wavelengths less than or equal to 1150 nm; C_C = $10^{[0.018(\lambda - 1150)]}$ for wavelengths greater than 1150 nm and less than 1200 nm; C_C = 8.0 + $10^{[0.04(\lambda - 1150)]}$ from 1250 to 1400 nm.

T_1 = 10 s for λ = 400 to 450 nm; T_1 = $10 \times 10^{[0.02 (\lambda - 550)]}$ for λ = 450 to 500 nm; and T_1 = 10 s for λ = 500 to 700.

For intermediate or large sources (e.g., laser diode arrays) at wavelengths between 400 nm and 1400 nm, the intrabeam viewing TLVs® can be increased by correction factor C_E (use Table 3) provided that the angular subtense α of the source (measured at the viewer's eye) is greater than α_{min}. C_E depends on α as follows:

Angular Subtense	Source Size Designation	Correction Factor C_E
$\alpha \leq \alpha_{min}$	Small	C_E = 1
$\alpha_{min} < \alpha \leq \alpha_{max}$	Intermediate	$C_E = \alpha/\alpha_{min}$
$\alpha > \alpha_{max}$	Large	$C_E = \alpha_{max}/\alpha_{min} = 3.33$ for t $\leq$ 0.625 ms
		$= 133.33\ t^{1/2}$ for 0.625 ms < t < 0.25 s
		$= 66.7$ for t $\geq$ 0.25 s

The angle referred to as α_{max} corresponds to the point where the TLVs® may be expressed as a constant radiance and the last equation can be rewritten in terms of radiance L.

$$L_{TLV} = (3.81 \times 10^5) \times (TLV_{pt\ source})\ J/(cm^2\ sr)\ for\ t\ < 0.625\ \mu s\ for\ 400 < \lambda < 700\ nm$$

$$L_{TLV} = 7.6\ t^{1/4}\ J/(cm^2\ sr)\ for\ 0.625\ ms < t < 0.25\ s\ for\ 400 < \lambda < 700\ nm$$

$$L_{TLV} = 4.8\ W/(cm^2\ sr)\ for\ t > 100\ s\ for\ 400 < \lambda < 700\ nm$$

Figure 5 illustrates these TLVs® for large sources expressed in terms of radiance.

The measurement aperture should be placed at a distance of 100 mm or greater from the source. For large area irradiation, the reduced TLV® for skin exposure applies as noted in the footnote to "IRB & C," Table 4.

TLV®–PA

TABLE 3. TLVs® for Extended Source Laser Viewing Conditions

Spectral Region	Wavelength	Exposure, (t) Seconds	TLV®
Light	400 to 700 nm	10^{-13} to 23×10^{-9}	$5\ C_E \times 10^{-8}$ J/cm²
	400 to 700 nm	23×10^{-9} to 50×10^{-9}	$4\ C_E \times 10^{15}\ t^3$ J/cm²
	400 to 700 nm	50×10^{-9} to 18×10^{-6}	$5.0\ C_E \times 10^{-7}$ J/cm²
	400 to 700 nm	18×10^{-6} to 0.7	$1.8\ C_E\ t^{0.75} \times 10^{-3}$ J/cm²

Dual Limits for 400 to 600 nm visible laser exposure for t > 0.7 s

Photochemical

For $\alpha \le 11$ mrad, the MPE is expressed as irradiance and radiant exposure*

	400 to 600 nm	0.7 to 100	$C_B \times 10^{-2}$ J/cm²
	400 to 600 nm	100 to 3×10^4	$C_B \times 10^{-4}$ W/cm²

For $\alpha > 11$ mrad, the MPE is expressed as radiance and integrated radiance*

	400 to 600 nm	0.7 to 1×10^4	$100\ C_B$ J/(cm² sr)
	400 to 600 nm	1×10^4 to 3×10^4	$C_B \times 10^{-2}$ W/(cm² sr)

and

Thermal

	400 to 700 nm	0.7 to T_2	$1.8\ C_E\ t^{0.75} \times 10^{-3}$ J/cm²
	400 to 700 nm	T_2 to 3×10^4	$1.8\ C_E\ T_2^{-0.25} \times 10^{-3}$ W/cm²

TLV®-PA

TABLE 3 (con't). TLVs® for Extended Source Laser Viewing Conditions

IRA	700 to 1050 nm	10^{-13} to 5×10^{-13}	$5\ C_E \times 10^{-8}$ J/cm²
	700 to 1050 nm	5×10^{-13} to 23×10^{-9}	$5\ C_A\ C_E \times 10^{-8}$ J/cm²
	700 to 1050 nm	23×10^{-9} to 50×10^{-9}	$4\ C_A\ C_E \times 10^{-15}\ t^3\ C_A$ J/cm²
	700 to 1050 nm	50×10^{-9} to 18×10^{-6}	$5\ C_A\ C_E \times 10^{-7}$ J/cm²
	700 to 1050 nm	18×10^{-6} to T_2	$1.8\ C_A\ C_E\ t^{0.75} \times 10^{-3}$ J/cm²
	700 to 1050 nm	T_2 to 3×10^4	$1.8\ C_A\ C_E\ T_2^{-0.25} \times 10$ W/cm²
	1050 to 1400 nm	10^{-13} to 5×10^{-13}	$C_C\ C_E \times 10^{-7}$ J/cm²
	1050 to 1400 nm	5×10^{-13} to 25×10^{-9}	$5.0\ C_C\ C_E \times 10^{-7}$ J/cm²
	1050 to 1400 nm	25×10^{-9} to 50×10^{-9}	$4.0 \times 10^{15}\ t^3\ C_C\ C_E$ J/cm²
	1050 to 1400 nm	50×10^{-9} to 50×10^{-6}	$5.0\ C_C\ C_E \times 10^{-6}$ J/cm²
	1050 to 1400 nm	50×10^{-6} to T_2	$9.0\ C_C\ C_E\ T_2^{0.75} \times 10^{-3}$ J/cm²
	1050 to 1400 nm	T_2 to 3×10^4	$9.0\ C_C\ C_E\ T_2^{-0.25} \times 10^{-3}$ W/cm²

* For sources subtending an angle greater than 11 mrad, the limit may also be expressed as an integrated radiance. $L_p = 100\ C_B$ J/(cm² sr) for 0.7 s $\leq t < 10^4$ s and $L_e = C_B \times 10^{-2}$ W/(cm² sr) for $t \geq 10^4$ s as measured through a limiting cone angle γ.

TABLE 3 (con't). TLVs® for Extended Source Laser Viewing Conditions

These correspond to values of J/cm² for $10 \leq t < 100$ s and W/cm² for $t \geq 100$ s as measured through a limiting cone angle γ.

$\gamma = 11$ mrad for $0.7 \leq t < 100$ s

$\gamma = 1.1 \times t^{0.5}$ mrad for $100 \leq t < 10^4$ s

$\gamma = 110$ mrad for $10^4 \leq t < 3 \times 10^4$ s

$T_2 = 10 \times 10^{(\alpha - 1.5)/98.5}$ for α expressed in mrad for $\lambda = 400$ to 1400 nm.

For exposure duration "t", the angle α_{max} is defined as:

$\alpha_{max} = 5$ mrad for $t \leq$ to 0.625 ms

$\alpha_{max} = 200\ t^{0.5}$ mrad for 0.625 ms $< t < 0.25$ s

$\alpha_{max} = 100$ mrad for $t \geq 0.25$ s, and

$\alpha_{min} = 1.5$ mrad

TLV®–PA

Notes for Tables 2 and 3

"NTE": To protect the cornea and lens: Change the 1 J/cm² to this set of dual limits for wavelengths between 400 nm and 1.5 μm. The lower of the TLVs® from Table 2 or Table 3 and the following apply.

		NTE (Second of Dual Limits)†
400 to 1200 nm	10^{-9} to 10^{-7}	$6\,C_A \times 10^{-2}$ J/cm²
400 to 1200 nm	10^{-7} to 10	$3.3\,C_A\,t^{1/4}$ J/cm²
400 to 1200 nm	10 to 3×10^4	$0.6\,C_A$ W/cm²
1200 to 1400 nm	10^{-9} to 10^{-3}	$10^{0(1.4-\lambda)}$ J/cm²
1200 to 1400 nm	10^{-3} to 4.0	$0.3 \times 10^{0(1.4-\lambda)} + 0.56\,t^{0.25} - 0.1$ J/cm²
1200 to 1400 nm	4.0 to 10	$0.3 \times 10^{0(1.4-\lambda)} + 0.7$ J/cm²
1200 to 1500 nm	10 to 3×10^4	$0.3 \times 10^{0(1.4-\lambda)} + 0.1\,t - 0.3$ W/cm²

†These dual limits will rarely apply except for exposures of very large angular subtense α – at least for wavelengths less than 1200 nm.

TABLE 4. TLVs® for Skin Exposure from a Laser Beam

Spectral Region	Wavelength	Exposure, (t) Seconds	TLV®
UV[A]	180 nm to 400 nm	10^{-9} to 10^4	Same as Table 2
	400 nm to 1400 nm	10^{-9} to 10^{-7}	$2 \, C_A \times 10^{-2} \, J/cm^2$
Light & IRA	" "	10^{-7} to 10	$1.1 \, C_A \, \sqrt[4]{t} \, J/cm^2$
	" "	10 to 3×10^4	$0.2 \, C_A \, W/cm^2$
IRB & C[B]	1.401 to 10^3 μm	10^{-14} to 3×10^4	Same as Table 2

[A] Ozone (O_3) is produced in air by sources emitting ultraviolet (UV) radiation at wavelengths below 250 nm. Refer to Chemical Substances TLV® for ozone.

$$C_A = 1.0 \text{ for } \lambda = 400 - 700 \text{ nm; see Figure 2 for } \lambda = 700 \text{ to } 1400 \text{ nm}$$

[B] At wavelengths greater than 1400 nm, for beam cross-sectional areas exceeding 100 cm², the TLV® for exposure durations exceeding 10 seconds is:

$$TLV = (10,000/A_s) \, mW/cm^2$$

where A_s is the irradiated skin area for 100 to 1000 cm², and the TLV® is 10 mW/cm² for irradiated skin areas exceeding 1000 cm² and is 100 mW/cm² for irradiated skin areas less than 100 cm².

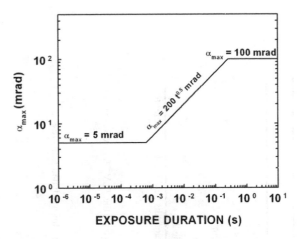

FIGURE 1. Variation of α_{max} with exposure duration.

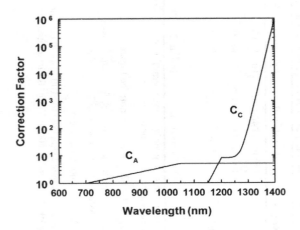

FIGURE 2. TLV® correction factors for $\lambda = 700–1400$ nm*
*For $\lambda = 700–1049$ nm; $C_A = 10^{(0.002[\lambda - 700])}$; for $\lambda = 1050–1400$ nm, $C_A = 5$.
For $\lambda \leq 1150$, $C_C = 1$; for $\lambda = 1150–1200$ nm, $C_C = 10^{[0.018(\lambda - 1150)]}$; and for $\lambda = 1200–1399$ nm, $C_C = 8 + 10^{[0.04(\lambda - 1250)]}$.

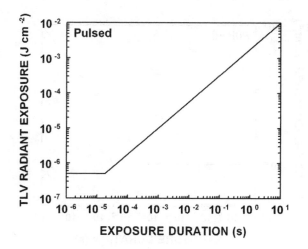

FIGURE 3a. TLV® for intrabeam (direct) viewing of laser beam (400–700 nm).

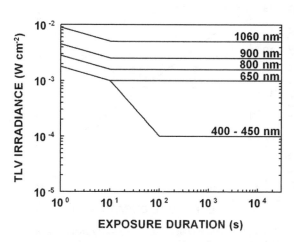

FIGURE 3b. TLV® for intrabeam (direct) viewing of CW laser beam (400–1400 nm).

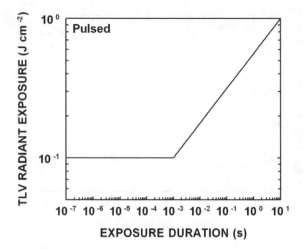

FIGURE 4a. TLV® for laser exposure of skin and eyes for far-infrared radiation (wavelengths greater than 1.4 μm).

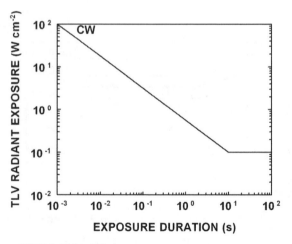

FIGURE 4b. TLV® for CW laser exposure of skin and eyes for far-infrared radiation (wavelengths greater than 1.4 μm).

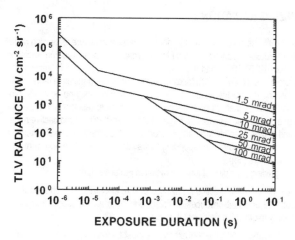

FIGURE 5. TLVs® in terms of radiance for exposures to extended-source lasers in the wavelength range of 400 to 700 nm.

* IONIZING RADIATION

ACGIH® has adopted as a TLV® for occupational exposure to ionizing radiation the guidelines recommended by the International Commission on Radiation Protection (ICRP, 2007) and the National Council on Radiation Protection and Measurements (NCRP, 1993). Ionizing radiation includes particulate radiation (α particles and β particles emitted from radioactive materials, and neutrons, protons and heavier charged particles produced in nuclear reactors and accelerators) and electromagnetic radiation (gamma rays emitted from radioactive materials and X-rays from electron accelerators and X-ray machines) with energy > 12.4 electron volts (eV) corresponding to wavelengths less than approximately 100 nanometers (nm).

The guiding principles of ionizing radiation protection are:

- **Justification:** No practice involving exposure to ionizing radiation should be adopted unless it produces sufficient benefit to an exposed individual or society to offset the detriment it causes.

- **Optimization:** All radiation exposures must be kept as low as reasonably achievable (ALARA), economic and social factors being taken into account.

- **Limitation:** The radiation dose from all occupationally relevant sources should not produce a level of risk of greater than about 10^{-3} per year of inducing fatal cancer during the lifetime of the exposed individual.

The TLV® guidelines are the dose limits shown in Table 1. Application of ALARA principles are recommended for all workers to keep radiation exposures as far below the guidelines as practicable.

TABLE 1. Guidelines for Exposure to Ionizing Radiation[A]

Type of Exposure	Dose Limits
Effective Dose;	
a) in any single year	50 mSv (millisievert)[B]
b) averaged over 5 years	20 mSv per year
Annual Equivalent Dose[C] to:	
a) lens of the eye	150 mSv
b) skin, hands and feet	500 mSv
Cumulative Effective Dose:	10 mSv × age in years
Embryo/Fetus Monthly Equivalent Dose[C]:	0.5 mSv
Radon and Radon Daughters	4 Working Level Months (WLM)[D]

[A] Doses are the effective doses from combined external and internal sources (excluding background radiation from radon, terrestrial, cosmic and internal body sources). The effective dose is that defined by ICRP and NCRP, where the effective dose is $H_T = \sum w_T \sum w_R D_{T,R}$, in which $D_{T,R}$ is the average absorbed dose in each tissue or organ, w_T is the tissue weighting factor representing the proportionate detriment (stochastic cancer risk), and w_R is the radiation weighting factor for the types of radiation(s) impinging on the body or, in the case of internal emitters, the radiation emitted by the source(s). The values of w_R and w_T to be used are those recommended by ICRP (2007).

[B] 10 mSv = 1 rem.

[C] The equivalent dose is the sum of external and internal absorbed doses multiplied by the appropriate radiation weighting factors.

[D] One WLM = 3.5×10^{-3} Jh/m³. The upper value for the individual worker annual dose is 10 mSv, which corresponds to an upper activity reference level of 1500 becquerels per m³ for radon and radon progeny in equilibrium, where a becquerel is a reciprocal second (ICRP, 1993, 2007).

TLV®-PA

ERGONOMICS

Ergonomics is the term applied to the field that studies and designs the human–machine interface to prevent illness and injury and to improve work performance. It attempts to ensure that jobs and work tasks are designed to be compatible with the capabilities of the workers. ACGIH® recognizes that some physical agents play an important role in ergonomics. Force and acceleration are addressed, in part, in the Hand–Arm Vibration (HAV) and Whole-Body Vibration (WBV) TLVs®. Thermal factors are addressed, in part, in the TLVs® for Thermal Stress. Force is also an important causal agent in injuries from lifting. Other important ergonomic considerations include work duration, repetition, contact stresses, postures, and psychosocial issues.

STATEMENT ON WORK-RELATED MUSCULOSKELETAL DISORDERS

ACGIH® recognizes work-related musculoskeletal disorders (MSDs) as an important occupational health problem that can be managed using an ergonomics health and safety program. The term musculoskeletal disorders refers to chronic muscle, tendon, and nerve disorders caused by repetitive exertions, rapid motions, high forces, contact stresses, extreme postures, vibration, and/or low temperatures. Other commonly used terms for work-related musculoskeletal disorders include cumulative trauma disorders (CTDs), repetitive motion illnesses (RMIs), and repetitive strain injuries (RSIs). Some of these disorders fit established diagnostic criteria such as carpal tunnel syndrome or tendinitis. Other musculoskeletal disorders may be manifested by nonspecific pain. Some transient discomfort is a normal consequence of work and is unavoidable, but discomfort that persists from day to day or interferes with activities of work or daily living should not be considered an acceptable outcome of work.

Control Strategies

The incidence and severity of MSDs are best controlled by an integrated ergonomics program. Major program elements include:
- Recognition of the problem,
- Evaluation of suspected jobs for possible risk factors,
- Identification and evaluation of causative factors,
- Involvement of workers as fully informed active participants, and
- Appropriate health care for workers who have developed musculoskeletal disorders.

General programmatic controls should be implemented when risk of MSDs is recognized. These include:
- Education of workers, supervisors, engineers, and managers;
- Early reporting of symptoms by workers; and
- Ongoing surveillance and evaluation of injury, health and medical data.

TLV®–PA

Job-specific controls are directed to individual jobs associated with MSDs. These include engineering controls and administrative controls. Personal protection may be appropriate under some limited circumstances.

Among engineering controls to eliminate or reduce risk factors from the job, the following may be considered:

- Using work methods engineering, e.g., time study, motion analysis, to eliminate unnecessary motions and exertions.
- Using mechanical assists to eliminate or reduce exertions required to hold tools and work objects.
- Selecting or designing tools that reduce force requirements, reduce holding time, and improve postures.
- Providing user-adjustable workstations that reduce reaching and improve postures.
- Implementing quality control and maintenance programs that reduce unnecessary forces and exertions, especially associated with nonvalue-added work.

Administrative controls reduce risk through reduction of exposure time and sharing the exposure among a larger group of workers. Examples include:

- Implementing work standards that permit workers to pause or stretch as necessary but at least once per hour.
- Re-allocating work assignments (e.g., using worker rotation or work enlargement) so that a worker does not spend an entire work shift performing high-demand tasks.

Due to the complex nature of musculoskeletal disorders, there is no "one size fits all" approach to reducing the incidence and severity of cases. The following principles apply to selecting actions:

- Appropriate engineering and administrative controls will vary from industry to industry and company to company.
- Informed professional judgment is required to select the appropriate control measures.
- Work-related MSDs typically require periods of weeks to months for recovery. Control measures should be evaluated accordingly to determine their effectiveness.

Nonoccupational Factors

It is not possible to eliminate all musculoskeletal disorders via engineering and administrative controls. There are individual and organizational factors that may influence the likelihood that an individual will experience musculoskeletal disorders. Some cases may be associated with nonoccupational factors such as:

- Rheumatoid arthritis
- Endocrinological disorders
- Acute trauma
- Obesity
- Pregnancy
- Age
- Gender

- Level of physical condition
- Previous injuries
- Diabetes
- Recreational/leisure activities

The recommended TLV® may not provide protection for people with these conditions and/or exposures. Engineering and administrative actions can help eliminate ergonomic barriers for persons with predisposing conditions and thus help to minimize disability.

Chronology of the Statement

1995: *Proposed* "Lifting Statement"
1996: Adopted with name change to "Musculoskeletal Statement"
2000: Editorial changes
2004: Editorial changes

HAND ACTIVITY LEVEL

Although work-related musculoskeletal disorders can occur in a number of body regions (including the shoulders, neck, low back, and lower extremities), the focus of this TLV® is on the hand, wrist, and forearm.

The TLV® shown in Figure 1 is based on epidemiological, psychophysical, and biomechanical studies and is intended for "mono-task" jobs performed for four or more hours per day. A mono-task job involves performing a similar set of motions or exertions repeatedly, such as working on an assembly line or using a keyboard and mouse. The TLV® specifically considers average hand activity level or "HAL" and peak hand force and represents conditions to which it is believed nearly all workers may be repeatedly exposed without adverse health effects.

HAL is based on the frequency of hand exertions and the duty cycle (distribution of work and recovery periods). HAL can be determined by trained observers based on exertion frequency, rest pauses and speed of motion using the rating scale shown in Figure 2. HAL also can be calculated from an analysis of the work method, force, and posture using information on hand exertion frequency and on duty cycle (work time/(work + rest time)) x 100% as described in Table 1 and in the *Documentation*.

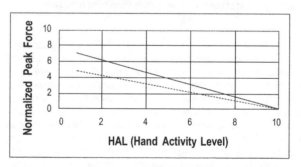

FIGURE 1. The TLV® for reduction of work-related musculoskeletal disorders based on "hand activity" or "HAL" and peak hand force. The top line depicts the TLV®. The bottom line is an Action Limit for which general controls are recommended.

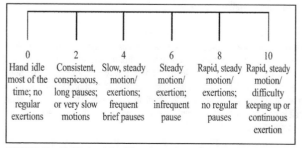

FIGURE 2. Hand Activity Level (0 to 10) can be rated using the above guidelines.

TABLE 1. Hand Activity Level (0 to 10) is Related to Exertion Frequency and Duty Cycle (% of work cycle where force is greater than 5% of maximum)

Frequency (exertion/s)	Period (s/exertion)	Duty Cycle (%)				
		0–20	20–40	40–60	60–80	80–100
0.125	8.0	1	1	—	—	—
0.25	4.0	2	2	3	—	—
0.5	2.0	3	4	5	5	6
1.0	1.0	4	5	5	6	7
2.0	0.5	—	5	6	7	8

Notes:
1. Round HAL values to the nearest whole number.
2. Use Figure 2 to obtain HAL values outside those listed in the table.

TLV®-PA

Peak hand force is the peak force exerted by the hand during each regular work cycle. Peak force can be determined with ratings by a trained observer, rated by workers using a Borg-like scale (see TLV® *Documentation* for definition), or measured using instrumentation, e.g., strain gauges or electromyography. In some cases, it can be calculated using biomechanical methods. These methods are intended to measure recurring peak forces; random force peaks associated with noise that occur less than 10% of the time are disregarded. Peak hand force is normalized on a scale of 0 to 10, which corresponds to 0% to 100% of the posture specific strength for the applicable population (males, females, young, old, office workers, factory workers, etc.):

Normalized Peak Force = (Peak force / Posture specific referent strength) × 10

The solid line in Figure 1 represents those combinations of force and hand activity level associated with a significantly elevated prevalence of musculoskeletal disorders. Appropriate control measures should be utilized so that the force for a given level of hand activity is below the upper solid line in Figure 1. It is not possible to specify a TLV® that protects all workers in all situations without profoundly affecting work rates. Therefore, an action limit is prescribed at which point general controls, including surveillance, are recommended.

Examples

1. Select a period of the job that represents an average activity. The selected period should include several complete work cycles. Videotapes may be used for documentation purposes and to facilitate rating of the job by others.
2. Rate the Hand Activity Level using the scale shown in Figure 2. Independent rating of jobs and discussion of results by three or more people can help produce a more precise rating than individual ratings.
3. Observe the job to identify forceful exertions and corresponding postures. Evaluate postures and forces using observer ratings, worker ratings, biomechanical analysis, or instrumentation. Normalized peak force is the required peak force divided by the representative maximum force for the posture multiplied by 10.

Consideration of Other Factors

Professional judgment should be used to reduce exposures below the action limits recommended in the HAL TLVs® if one or more of the following factors are present:

- sustained non-neutral postures such as wrist flexion, extension, wrist deviation, or forearm rotation;
- contact stresses;
- low temperatures; or
- vibration.

Employ appropriate control measures any time the TLV® is exceeded or an elevated incidence of work-related musculoskeletal disorders is detected.

TLV®-PA

LIFTING

These TLVs® recommend workplace lifting conditions under which it is believed nearly all workers may be repeatedly exposed, day after day, without developing work-related low back and shoulder disorders associated with repetitive lifting tasks. There are individual and organizational risk factors that may influence the likelihood that an individual will experience low back and shoulder disorders.

Lifting TLVs®

The TLVs® consist of three tables with weight limits, in kilograms (kg), for two-handed, mono-lifting tasks within 30 degrees of the sagittal [neutral] plane. A mono-lifting task is one in which the loads are similar and the starting and destination points are repeated, and this is the only lifting task performed during the day. Other manual material-handling tasks such as carrying, pushing, and pulling are not accounted for in the TLV®, and care must be exercised in applying the TLVs® under these circumstances.

These TLVs® (Tables 1 through 3) are presented for lifting tasks defined by their durations, either less than or greater than 2 hours per day, and by their frequency, expressed in number of lifts per hour, as qualified in the *Notes* to each table.

In the presence of any factor(s) or working condition(s) listed below, professional judgment should be used to reduce weight limits below those recommended in the TLVs®:

- High-frequency lifting: > 360 lifts per hour.
- Extended work shifts: lifting performed for longer than 8 hours per day.
- High asymmetry: lifting more than 30 degrees away from the sagittal plane.
- Rapid lifting motions and motions with twisting (e.g., from side to side).
- One-handed lifting.
- Constrained lower body posture, such as lifting while seated or kneeling.
- High heat and humidity (*see* Heat Stress and Heat Strain TLVs®).
- Lifting unstable objects (e.g., liquids with shifting center of mass or lack of coordination or equal sharing in multi-person lifts).
- Poor hand coupling: lack of handles, cut-outs, or other grasping points.
- Unstable footing (e.g., inability to support the body with both feet while standing).
- During or immediately after exposure to whole-body vibration at or above the TLV® for Whole-Body Vibration (*see* the current *TLV® Documentation* for Whole-Body Vibration).

Instructions for Users

1. **Read the *Documentation*** for the Lifting TLVs® so you understand the basis for these TLVs® and their limitations.
2. **Classify task duration** as less than or equal to a cumulative 2 hours per day or greater than a cumulative 2 hours per day. Task duration is the total length of time that a worker performs the task in 1 day.

TABLE 1. TLVs® for Lifting Tasks:
≤ 2 Hours per Day with ≤ 60 Lifts per Hour
OR
>2 Hours per Day with ≤ 12 Lifts per Hour

	Horizontal Zone[A]		
Vertical Zone	**Close:** < 30 cm	**Inter-mediate:** 30 to 60 cm	**Extended:[B]** > 60 to 80 cm
Reach limit[C] or 30 cm above shoulder to 8 cm below shoulder height	16 kg	7 kg	No known safe limit for repetitive lifting[D]
Knuckle height[E] to below shoulder	32 kg	16 kg	9 kg
Middle shin to knuckle height[E]	18 kg	14 kg	7 kg
Floor to middle shin height	14 kg	No known safe limit for repetitive lifting[D]	No known safe limit for repetitive lifting[D]

Footnotes for Tables 1 through 3:

A. Distance from midpoint between inner ankle bones and the load.
B. Lifting tasks should not start or end at a horizontal reach distance more than 80 cm from the midpoint between the inner ankle bones (Figure 1).
C. Routine lifting tasks should not start or end at heights that are greater than 30 cm above the shoulder or more than 180 cm above floor level (Figure 1).
D. Routine lifting tasks should not be performed for shaded table entries marked "No known safe limit for repetitive lifting." While the available evidence does not permit identification of safe weight limits in the shaded regions, professional judgment may be used to determine if infrequent lifts of light weights may be safe.
E. Anatomical landmark for knuckle height assumes the worker is standing erect with arms hanging at the sides.

TLV®-PA

TABLE 2. TLVs® for Lifting Tasks
>2 Hours per Day with > 12 and ≤ 30 Lifts per Hour
OR
≤ 2 Hours per Day with > 60 and ≤ 360 Lifts per Hour

Vertical Zone	Horizontal Zone[A]		
	Close: < 30 cm	Inter-mediate: 30 to 60 cm	Extended:[B] > 60 to 80 cm
Reach limit[C] or 30 cm above shoulder to 8 cm below shoulder height	14 kg	5 kg	No known safe limit for repetitive lifting[D]
Knuckle height[E] to below shoulder	27 kg	14 kg	7 kg
Middle shin to knuckle height[E]	16 kg	11 kg	5 kg
Floor to middle shin height	9 kg	No known safe limit for repetitive lifting[D]	No known safe limit for repetitive lifting[D]

See Notes in Table 1.

TABLE 3. TLVs® for Lifting Tasks
>2 Hours per Day with > 30 and ≤ 360 Lifts per Hour

Vertical Zone	Horizontal Zone[A]		
	Close: < 30 cm	Inter-mediate: 30 to 60 cm	Extended:[B] > 60 to 80 cm
Reach limit[C] from 30 cm above to 8 cm below shoulder height	11 kg	No known safe limit for repetitive lifting[D]	No known safe limit for repetitive lifting[D]
Knuckle height[E] to below shoulder	14 kg	9 kg	5 kg
Middle shin to knuckle height[E]	9 kg	7 kg	2 kg
Floor to middle shin height	No known safe limit for repetitive lifting[D]	No known safe limit for repetitive lifting[D]	No known safe limit for repetitive lifting[D]

See Notes in Table 1.

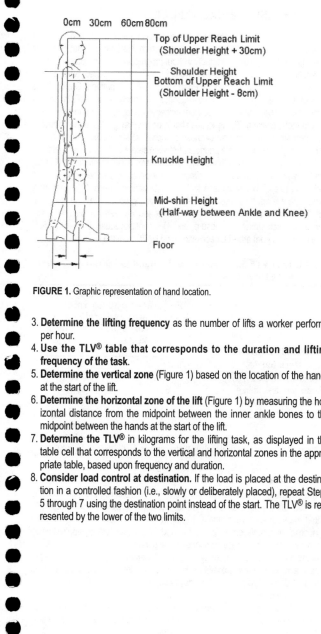

0cm 30cm 60cm 80cm

Top of Upper Reach Limit
(Shoulder Height + 30cm)

Shoulder Height
Bottom of Upper Reach Limit
(Shoulder Height - 8cm)

Knuckle Height

Mid-shin Height
(Half-way between Ankle and Knee)

Floor

FIGURE 1. Graphic representation of hand location.

3. **Determine the lifting frequency** as the number of lifts a worker performs per hour.
4. **Use the TLV® table that corresponds to the duration and lifting frequency of the task**.
5. **Determine the vertical zone** (Figure 1) based on the location of the hands at the start of the lift.
6. **Determine the horizontal zone of the lift** (Figure 1) by measuring the horizontal distance from the midpoint between the inner ankle bones to the midpoint between the hands at the start of the lift.
7. **Determine the TLV®** in kilograms for the lifting task, as displayed in the table cell that corresponds to the vertical and horizontal zones in the appropriate table, based upon frequency and duration.
8. **Consider load control at destination.** If the load is placed at the destination in a controlled fashion (i.e., slowly or deliberately placed), repeat Steps 5 through 7 using the destination point instead of the start. The TLV® is represented by the lower of the two limits.

HAND–ARM (SEGMENTAL) VIBRATION

The TLVs® in Table 1 refer to component acceleration levels and durations of exposure that represent conditions under which it is believed that nearly all workers may be exposed repeatedly without progressing beyond Stage 1 of the Stockholm Workshop Classification System for Vibration-induced White Finger (VWF), also known as Raynaud's Phenomenon of Occupational Origin (Table 2). Since there is a paucity of dose–response relationships for VWF, these recommendations have been derived from epidemiological data from forestry, mining, and metal working. These values should be used as guides in the control of hand–arm vibration exposure; because of individual susceptibility, they should not be regarded as defining a boundary between safe and dangerous levels.

It should be recognized that control of hand–arm vibration syndrome (HAVS) from the workplace cannot occur simply by specifying and adhering to a given TLV®. The use of 1) antivibration tools, 2) antivibration gloves, 3) proper work practices that keep the worker's hands and remaining body warm and also minimize the vibration coupling between the worker and the vibration tool are necessary to minimize vibration exposure, and 4) a conscientiously applied medical surveillance program are ALL necessary to rid HAVS from the workplace.

TABLE 1. TLVs® for Exposure of the Hand to Vibration in Either X_h, Y_h, or Z_h Directions

Total Daily Exposure Duration ☆	Values of the Dominant,★ Frequency-Weighted, rms, Component Acceleration Which Shall not be Exceeded $a_K,(a_{K_{eq}})$	
	m/s²	g△
4 hours and less than 8	4	0.40
2 hours and less than 4	6	0.61
1 hour and less than 2	8	0.81
less than 1 hour	12	1.22

☆ The total time vibration enters the hand per day, whether continuously or intermittently.

★ Usually one axis of vibration is dominant over the remaining two axes. If one or more vibration axes exceeds the Total Daily Exposure, then the TLV® has been exceeded.

g△ = 9.81 m/s².

Notes for Table 1:

1. The weighting network provided in Figure 1 is considered the best available to frequency weight acceleration components. However, studies suggest that the frequency weighting at higher frequencies (above 16 Hz) may not incorporate a sufficient safety factor, and CAUTION must be applied when tools with high-frequency components are used.

2. Acute exposures to frequency-weighted, root-mean-square (rms), component accelerations in excess of the TLVs® for infrequent periods of time (e.g., 1 day per week or several days over a 2-week period) are not necessarily more harmful.

3. Acute exposures to frequency-weighted, rms, component accelerations of three times the magnitude of the TLVs® are expected to result in the same health effects after 5 to 6 years of exposure.

TABLE 2. Stockholm Workshop HAVS Classification System for Cold-induced Peripheral Vascular and Sensorineural Symptoms

Vascular Assessment		
Stage	Grade	Description
0	—	No attacks
1	Mild	Occasional attacks affecting only the tips of one or more fingers
2	Moderate	Occasional attacks affecting distal and middle (rarely also proximal) phalanges of one or more fingers
3	Severe	Frequent attacks affecting ALL phalanges of most fingers
4	Very Severe	As in Stage 3, with trophic skin changes in the finger tips

Note: Separate staging is made for each hand, e.g., 2L(2)/1R(1) = stage 2 on left hand in two fingers: stage 1 on right hand in one finger.

- -

Sensorineural Assessment	
Stage	Symptoms
0SN	Exposed to vibration but no symptoms
1SN	Intermittent numbness, with or without tingling
2SN	Intermittent or persistent numbness, reducing sensory perception
3SN	Intermittent or persistent numbness, reducing tactile discrimination and/or manipulative dexterity

Note: Separate staging is made for each hand.

TLV®-PA

4. To moderate the adverse effects of vibration exposure, workers should be advised to avoid continuous vibration exposure by cessation of vibration exposure for approximately 10 minutes per continuous vibration hour.
5. Good work practices should be used and should include instructing workers to employ a minimum hand grip force consistent with safe operation of the power tool or process, to keep their body and hands warm and dry, to avoid smoking, and to use antivibration tools and gloves when possible. As a general rule, gloves are more effective for damping vibration at high frequencies.
6. A vibration measurement transducer, together with its device for attachment to the vibration source, should weigh less than 15 grams and should possess a cross-axis sensitivity of less than 10%.
7. The measurement by many (mechanically underdamped) piezoelectric accelerometers of repetitive, large displacement, impulsive vibrations, such as those produced by percussive pneumatic tools, is subject to error. The insertion of a suitable, low-pass, mechanical filter between the accelerometer and the source of vibration with a cut-off frequency of 1500 Hz or greater (and cross-axis sensitivity of less than 10%) can help eliminate incorrect readings.
8. The manufacturer and type number of all apparatus used to measure vibration should be reported, as well as the value of the dominant direction and frequency-weighted, rms, component acceleration.

Continuous, Intermittent, Impulsive, or Impact Hand–Arm Vibration

The measurement of vibration should be performed in accordance with the procedures and instrumentation specified by ISO 5349 (1986)[1] or ANSI S3.34-1986[2] and summarized below.

The acceleration of a vibration handle or work piece should be determined in three mutually orthogonal directions at a point close to where vibration enters the hand. The directions should preferably be those forming the biodynamic coordinate system but may be a closely related basicentric system with its origin at the interface between the hand and the vibrating surface (Figure 2) to accommodate different handle or work piece configurations. A small and lightweight transducer should be mounted so as to record accurately one or more orthogonal components of the source vibration in the frequency range from 5 to 1500 Hz. Each component should be frequency-weighted by a filter network with gain characteristics specified for human-response vibration measuring instrumentation, to account for the change in vibration hazard with frequency (Figure 1).

Assessment of vibration exposure should be made for EACH applicable direction (X_h, Y_h, Z_h) since vibration is a vector quantity (magnitude and direction). In each direction, the magnitude of the vibration during normal operation of the power tool, machine, or work piece should be expressed by the root-mean-square (rms) value of the frequency-weighted component accelerations, in units of meters per second squared (m/s^2), or gravitational units (g), the largest of which, a_K, forms the basis for exposure assessment.

For each direction being measured, linear integration should be employed for vibrations that are of extremely short duration or vary substantially in time. If the total daily vibration exposure in a given direction is

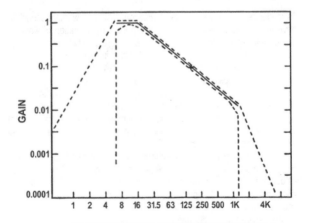

FIGURE 1. Gain characteristics on the filter network used to frequency-weight acceleration components (continuous line). The filter tolerances (dashed lines) are those contained in ISO 5349 and ANSI S3.34-1986.

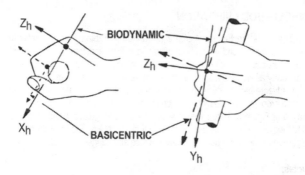

FIGURE 2. Biodynamic and basicentric coordinate systems for the hand, showing the directions of the acceleration components (ISO 5349[1] and ANSI S3.34–1986[2]).

composed of several exposures at different rms accelerations, then the equivalent, frequency-weighted component acceleration in that direction should be determined in accordance with the following equation:

$$\left(a_{K_{eq}}\right) = \left[\frac{1}{T} \sum_{i=1}^{n} \left(a_{K_i}\right)^2 T_i\right]^{1/2}$$

$$= \sqrt{\left(a_{K_1}\right)^2 \frac{T_1}{T} + \left(a_{K_2}\right)^2 \frac{T_2}{T} + \ldots \left(a_{K_n}\right)^2 \frac{T_n}{T}}$$

where: $T = \sum_{i=1}^{n} T_i$

$T =$ total daily exposure duration
$a_{K_i} =$ *ith* frequency-weighted, rms acceleration component with duration T_i

These computations may be performed by commercially available human-response vibration measuring instruments.

References

1. International Standards Organization: ISO 5349 (1986): Guide for the Measurement and the Assessment of Human Exposure to Hand Transmitted Vibration. ISO, Geneva (1986).
2. American National Standards Institute: ANSI S3.34-1986: Guide for the Measurement and Evaluation of Human Exposure to Vibration Transmitted to the Hand. ANSI, New York (1986).

TLV®-PA

WHOLE–BODY VIBRATION

The TLVs® in Figures 1 and 2 (tabulated in Tables 1 and 2) refer to mechanically induced whole-body vibration (WBV) acceleration component root-mean-square (rms) magnitudes and durations under which it is believed that nearly all workers may be exposed repeatedly with minimum risk of back pain, adverse health effects to the back, and inability to operate a land-based vehicle properly. The biodynamic coordinate system to which they apply is displayed in Figure 3. These values should be used as guides in the control of WBV exposure, but because of individual susceptibility, they should not be regarded as defining a boundary between safe and dangerous levels.

Notes:

1. Vibration acceleration is a vector with magnitude expressed in units of m/s^2. The gravitational acceleration, g, equals 9.81 m/s^2.

2. Figures 1 and 2 each show a family of daily exposure time–dependent curves. They indicate that human vibration resonance occurs in the 4 to 8 Hz frequency range for the z axis and in the 1 to 2 Hz frequency range for the x and y axes, where the axes are defined in Figure 3.

3. WBV measurements and equivalent exposure time calculations for interrupted exposures, where the rms acceleration levels vary appreciably over time, should be made according to ISO 2631 or ANSI S3.18-1979.[1,2]

4. The TLV® is valid for vibration crest factors of 6 or less. Crest factor is defined as the ratio of peak to rms acceleration, measured in the same direction, over a period of 1 minute for any of the orthogonal x, y, and z axes. The TLV® will underestimate the effects of WBV and must be used with caution when the crest factor exceeds 6.

5. The TLV® is not intended for use in fixed buildings (see ANSI S3.29-1983),[3] in off-shore structures, or in ships.

6. A summary of WBV measurement and data analysis procedures follows:[4]

    a. At each measurement point, three orthogonal, continuous, rms acceleration measurements are simultaneously made and recorded for at least 1 minute along the biodynamic coordinates shown in Figure 3.

    b. Three very light-weight accelerometers, each with a cross-axis sensitivity of less than 10%, are perpendicularly mounted to a light-weight metal cube and placed in the center of a hard rubber disc (per SAE, J1013).[5] The total weight of the disc, cube, accelerometers, and cables should not exceed 10% of the total weight of the object to be measured. Measurements are made by placing the instrumented rubber disc on the top of the driver's seat, under the driver's buttocks, as the vehicle is operated.

    c. For each axis, a $^1/_3$ octave band (1 to 80 Hz), separate Fourier spectrum analysis is required for comparison to Figure 1 or Figure 2, as appropriate.

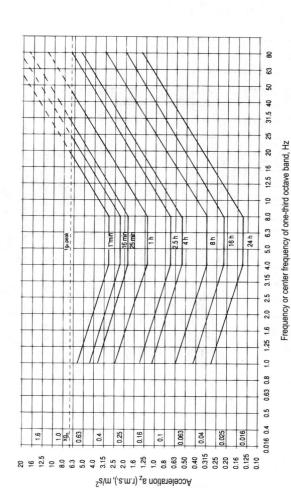

FIGURE 1. Longitudinal (a_z) acceleration TLVs® as a function of frequency and exposure time. Adapted from ISO 2631.[1]

TLV®–PA

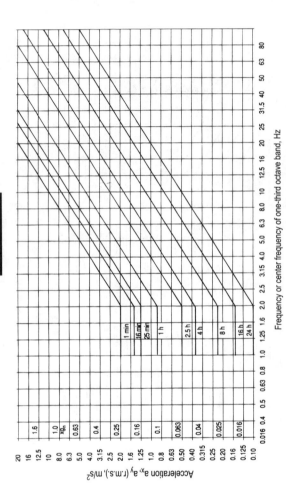

FIGURE 2. Transverse (a_x, a_y) acceleration TLVs® as a function of frequency and exposure time. Adapted from ISO 2631.[1]

TABLE 1. Numerical values for vibration acceleration in the longitudinal, a_z, direction [foot-to-head direction] [see Figure 1]. Values define the TLV® in terms of rms value of pure (sinusoidal) single-frequency vibration or of rms value in one-third-octave band for distributed vibration. (Adapted from ISO 2631)

Acceleration, m/s²

Frequency Hz	Exposure times 24 h	16 h	8 h	4 h	2.5 h	1 h	25 min	16 min	1 min
1.0	0.280	0.383	0.63	1.06	1.40	2.36	3.55	4.25	5.60
1.25	0.250	0.338	0.56	0.95	1.26	2.12	3.15	3.75	5.00
1.6	0.224	0.302	0.50	0.85	1.12	1.90	2.80	3.35	4.50
2.0	0.200	0.270	0.45	0.75	1.00	1.70	2.50	3.00	4.00
2.5	0.180	0.239	0.40	0.67	0.90	1.50	2.24	2.65	3.55
3.15	0.160	0.212	0.355	0.60	0.80	1.32	2.00	2.35	3.15
4.0	0.140	0.192	0.315	0.53	0.71	1.18	1.80	2.12	2.80
5.0	0.140	0.192	0.315	0.53	0.71	1.18	1.80	2.12	2.80
6.3	0.140	0.192	0.315	0.53	0.71	1.18	1.80	2.12	2.80
8.0	0.140	0.192	0.315	0.53	0.71	1.18	1.80	2.12	2.80
10.0	0.180	0.239	0.40	0.67	0.90	1.50	2.24	2.65	3.55
12.5	0.224	0.302	0.50	0.85	1.12	1.90	2.80	3.35	4.50
16.0	0.280	0.383	0.63	1.06	1.40	2.36	3.55	4.25	5.60
20.0	0.355	0.477	0.80	1.32	1.80	3.00	4.50	5.30	7.10
25.0	0.450	0.605	1.0	1.70	2.24	3.75	5.60	6.70	9.00
31.5	0.560	0.765	1.25	2.12	2.80	4.75	7.10	8.50	11.2
40.0	0.710	0.955	1.60	2.65	3.55	6.00	9.00	10.6	14.0
50.0	0.900	1.19	2.0	3.35	4.50	7.50	11.2	13.2	18.0
63.0	1.120	1.53	2.5	4.25	5.60	9.50	14.0	17.0	22.4
80.0	1.400	1.91	3.15	5.30	7.10	11.8	18.0	21.2	28.0

TLV®–PA

TLV®–PA

TABLE 2. Numerical values for vibration acceleration in the transverse, a_x or a_y, direction [back-to-chest or side-to-side] [see Figure 2]. Values define the TLV® in terms of rms value of pure (sinusoidal) single-frequency vibration or of rms value in one-third-octave band for distributed vibration. (Adapted from ISO 2631)

| Frequency Hz | Acceleration, m/s² | | | | | | | | |
| | Exposure times | | | | | | | | |
	24 h	16 h	8 h	4 h	2.5 h	1 h	25 min	16 min	1 min
1.0	0.100	0.135	0.224	0.355	0.50	0.85	1.25	1.50	2.0
1.25	0.100	0.135	0.224	0.355	0.50	0.85	1.25	1.50	2.0
1.6	0.100	0.135	0.224	0.355	0.50	0.85	1.25	1.50	2.0
2.0	0.100	0.135	0.224	0.355	0.50	0.85	1.25	1.50	2.0
2.5	0.125	0.171	0.280	0.450	0.63	1.06	1.6	1.9	2.5
3.15	0.160	0.212	0.355	0.560	0.8	1.32	2.0	2.36	3.15
4.0	0.200	0.270	0.450	0.710	1.0	1.70	2.5	3.0	4.0
5.0	0.250	0.338	0.560	0.900	1.25	2.12	3.15	3.75	5.0
6.3	0.315	0.428	0.710	1.12	1.6	2.65	4.0	4.75	6.3
8.0	0.40	0.54	0.900	1.40	2.0	3.35	5.0	6.0	8.0
10.0	0.50	0.675	1.12	1.80	2.5	4.25	6.3	7.5	10.0
12.5	0.63	0.855	1.40	2.24	3.15	5.30	8.0	9.5	12.5
16.0	0.80	1.06	1.80	2.80	4.0	6.70	10.0	11.8	16.0
20.0	1.00	1.35	2.24	3.55	5.0	8.5	12.5	15.0	20.0
25.0	1.25	1.71	2.80	4.50	6.3	10.6	15.0	19.0	25.0
31.5	1.60	2.12	3.55	5.60	8.0	13.2	20.0	23.6	31.5
40.0	2.00	2.70	4.50	7.10	10.0	17.0	25.0	30.0	40.0
50.0	2.50	3.38	5.60	9.00	12.5	21.2	31.5	37.5	50.0
63.0	3.15	4.28	7.10	11.2	16.0	26.5	40.0	45.7	63.0
80.0	4.00	5.4	9.00	14.0	20.0	33.5	50.0	60.0	80.0

    d. If the rms acceleration of any of the spectral peaks equals or exceeds the values shown in Figure 1 or Figure 2 for the relevant time periods, then the TLV® is exceeded for that exposure time. The axis with the highest spectral peak intersecting the curve with the shortest exposure time dominates and determines the permissible exposure.

7. The total-weighted rms acceleration for each axis can be calculated using Equation 1 with the appropriate axis weighting factors taken from Table 3. For the x axis (analogous equations and definitions apply to the y and z axes), the equation is:

$$A_{wx} = \sqrt{\Sigma \left(W_{fx} A_{fx} \right)^2} \qquad (1)$$

where: A_{wx} = total weighted rms acceleration for the x axis
                W_{fx} = weighting factor for the x axis at each ⅓ octave band frequency from 1 to 80 Hz (Table 3)
                A_{fx} = rms acceleration value for the x axis spectrum at each ⅓ octave band frequency from 1 to 80 Hz

8. If the vibration axes have similar acceleration magnitudes as determined by Equation 1, the combined motion of all three axes could be greater than any one component and could possibly affect vehicle operator performance.[1,2] Each of the component results determined by Equation 1 may be

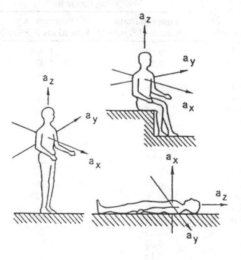

FIGURE 3. Biodynamic coordinate system acceleration measurements (adapted from ISO 2631). a_x, a_y, a_z = acceleration in the direction of the x, y, and z axes; x axis = back-to-chest; y axis = right-to-left; z axis = foot-to-head.

used in Equation 2 to find the resultant, which is the overall weighted total rms acceleration, A_{wt}:

$$A_{wt} = \sqrt{(1.4\,A_{wx})^2 + (1.4\,A_{wy})^2 + (A_{wz})^2} \qquad (2)$$

The factor of 1.4 multiplying the x and y total, weighted rms acceleration values is the ratio of the values of the longitudinal and transverse curves of equal response in the most sensitive human response ranges.

The Commission of the European Communities now recommends 0.5 m/s² as an action level for an 8 hour per day overall weighted total rms acceleration. This may be compared with the results of Equation 2.

9. Short-duration, high-amplitude, multiple-vibration shocks may occur with crest factors greater than 6 during the workday, in which cases the TLV® may not be protective (Note 4). Other methods of calculation that include the "4th power concept" may be desirable in these instances.[6]

TABLE 3. Weighting Factors Relative to the Frequency Range of Maximum Acceleration Sensitivity[A] for the Response Curves of Figures 1 and 2 (Adapted from ISO 2631)

	Weighting factor for	
Frequency Hz	Longitudinal z Vibrations [Figure 1]	Transverse x,y Vibrations [Figure 2]
1.0	0.50	1.00
1.25	0.56	1.00
1.6	0.63	1.00
2.0	0.71	1.00
2.5	0.80	0.80
3.15	0.90	0.63
4.0	1.00	0.5
5.0	1.00	0.4
6.3	1.00	0.315
8.0	1.00	0.25
10.0	0.80	0.2
12.5	0.63	0.16
16.0	0.50	0.125
20.0	0.40	0.1
25.0	0.315	0.08
31.5	0.25	0.063
40.0	0.20	0.05
50.0	0.16	0.04
63.0	0.125	0.0315
80.0	0.10	0.025

[A] 4 to 8 Hz in the case of $\pm\,a_z$ resonance vibration.
1 to 2 Hz in the case of $\pm\,a_y$ or a_x resonance vibration.

TLV®-PA

10. WBV controls may include the use of "air-ride" suspended seats, suspended cabs, maintenance of vehicle suspension systems, proper tire inflation, and remote control of vibrating processes. Seats with arm rests, lumbar support, an adjustable seat back, and an adjustable seat pan are also useful.

11. The following good work practices may also be useful for workers operating vehicles:[7,8]
    a. Avoid lifting or bending immediately following exposure.
    b. Use simple motions, with minimum rotation or twisting, when exiting a vehicle.

References

1. International Standards Organization: ISO 2631/1: Evaluation of Human Exposure to Whole-Body Vibration. ISO, Geneva (1985).
2. American National Standards Institute: ANSI S3.18: Guide for the Evaluation of Human Exposure to Whole-Body Vibration. ANSI, New York (1979).
3. American National Standards Institute: ANSI S3.29: Guide for the Evaluation of Human Exposure to Whole-Body Vibration in Buildings. ANSI, New York (1983).
4. Wasserman, D: Human Aspects of Occupational Vibration. Elsevier Publishers, Amsterdam (1987).
5. Society of Automotive Engineers. SAE J1013: Measurement of Whole Body Vibration of the Seated Operator of Off Highway Work Machines. SAE, Warrendale, PA (August 1992).
6. Griffin, M: Handbook of Human Vibration. Academic Press, London (1990).
7. Wilder, D: The Biomechanics of Vibration and Low Back Pain. Am. J. Ind. Med. 23:577–588 (1993).
8. Wilder, D; Pope, M; Frymoyer, J: The Biomechanics of Lumbar Disc Herniation and the Effect of Overload and Instability. J. Spinal Disorders 1:16–32 (1988).

THERMAL STRESS

COLD STRESS

The cold stress TLVs® are intended to protect workers from the severest effects of cold stress (hypothermia) and cold injury and to describe exposures to cold working conditions under which it is believed nearly all workers can be repeatedly exposed without adverse health effects. The TLV® objective is to prevent the deep body temperature from falling below 36°C (96.8°F) and to prevent cold injury to body extremities (deep body temperature is the core temperature of the body determined by conventional methods for rectal temperature measurements). For a single, occasional exposure to a cold environment, a drop in core temperature to no lower that 35°C (95°F) should be permitted. In addition to provisions for total body protection, the TLV® objective is to protect all parts of the body with emphasis on hands, feet, and head from cold injury.

Introduction

Fatal exposures to cold among workers have almost always resulted from accidental exposures involving failure to escape from low environmental air temperatures or from immersion in low temperature water. The single most important aspect of life-threatening hypothermia is the fall in the deep core temperature of the body. The clinical presentations of victims of hypothermia are shown in Table 1. Workers should be protected from exposure to cold so that the deep core temperature does not fall below 36°C (96.8°F); lower body temperatures will very likely result in reduced mental alertness, reduction in rational decision making, or loss of consciousness with the threat of fatal consequences.

Pain in the extremities may be the first early warning of danger to cold stress. During exposure to cold, maximum severe shivering develops when the body temperature has fallen to 35°C (95°F). This must be taken as a sign of danger to the workers and exposure to cold should be immediately terminated for any workers when severe shivering becomes evident. Useful physical or mental work is limited when severe shivering occurs.

Since prolonged exposure to cold air, or to immersion in cold water, at temperatures well above freezing can lead to dangerous hypothermia, whole body protection must be provided.

1. Adequate insulating dry clothing to maintain core temperatures above 36°C (96.8°F) must be provided to workers if work is performed in air temperatures below 4°C (40°F). Wind chill cooling rate and the cooling power of air are critical factors. [Wind chill cooling rate is defined as heat loss from a body expressed in watts per meter squared which is a function of the air temperature and wind velocity upon the exposed body.] The higher the wind speed and the lower the temperature in the work area, the greater the insulation value of the protective clothing required. An equivalent chill temperature chart relating the actual dry bulb air temperature and the wind velocity is presented in Table 2. The equivalent chill temperature should beused when estimating the combined cooling effect of wind and low air temperatures on exposed skin or when determining clothing insulation requirements to maintain the deep body core temperature.

TABLE 1. Progressive Clinical Presentations of Hypothermia☆

Core Temperature		Clinical Signs
°C	°F	
37.6	99.6	"Normal" rectal temperature
37	98.6	"Normal" oral temperature
36	96.8	Metabolic rate increases in an attempt to compensate for heat loss
35	95.0	Maximum shivering
34	93.2	Victim conscious and responsive, with normal blood pressure
33	91.4	Severe hypothermia below this temperature
32⎱ 31⎰	89.6⎱ 87.8⎰	Consciousness clouded; blood pressure becomes difficult to obtain; pupils dilated but react to light; shivering ceases
30⎱ 29⎰	86.0⎱ 84.2⎰	Progressive loss of consciousness; muscular rigidity increases; pulse and blood pressure difficult to obtain; respiratory rate decreases
28	82.4	Ventricular fibrillation possible with myocardial irritability
27	80.6	Voluntary motion ceases; pupils nonreactive to light; deep tendon and superficial reflexes absent
26	78.8	Victim seldom conscious
25	77.0	Ventricular fibrillation may occur spontaneously
24	75.2	Pulmonary edema
22⎱ 21⎰	71.6⎱ 69.8⎰	Maximum risk of ventricular fibrillation
20	68.0	Cardiac standstill
18	64.4	Lowest accidental hypothermia victim to recover
1 7	62.6	Isoelectric electroencephalogram
9	48.2	Lowest artificially cooled hypothermia patient to recover

☆Presentations approximately related to core temperature. Reprinted from the January 1982 issue of *American Family Physician,* published by the American Academy of Family Physicians.

2. Unless there are unusual or extenuating circumstances, cold injury to other than hands, feet, and head is not likely to occur without the development of the initial signs of hypothermia. Older workers or workers with circulatory problems require special precautionary protection against cold injury. The use of extra insulating clothing and/or a reduction in the duration of the exposure period are among the special precautions which should be considered. The precautionary actions to be taken will depend upon the physical condition of the worker and should be determined with the advice of a physician with knowledge of the cold stress factors and the medical condition of the worker.

TLV®-PA

TLV®–PA

TABLE 2. Cooling Power of Wind on Exposed Flesh Expressed as Equivalent Temperature (under calm conditions) *

Estimated Wind Speed (in mph)	Actual Temperature Reading (°F)											
	50	40	30	20	10	0	-10	-20	-30	-40	-50	-60
	Equivalent Chill Temperature (°F)											
calm	50	40	30	20	10	0	-10	-20	-30	-40	-50	-60
5	48	37	27	16	6	-5	-15	-26	-36	-47	-57	-68
10	40	28	16	4	-9	-24	-33	-46	-58	-70	-83	-95
15	36	22	9	-5	-18	-32	-45	-58	-72	-85	-99	-112
20	32	18	4	-10	-25	-39	-53	-67	-82	-96	-110	-121
25	30	16	0	-15	-29	-44	-59	-74	-88	-104	-118	-133
30	28	13	-2	-18	-33	-48	-63	-79	-94	-109	-125	-140
35	27	11	-4	-20	-35	-51	-67	-82	-98	-113	-129	-145
40	26	10	-6	-21	-37	-53	-69	-85	-100	-116	-132	-148

(Wind speeds greater than 40 mph have little additional effect.)

LITTLE DANGER
In < hr with dry skin. Maximum danger of false sense of security.

INCREASING DANGER
Danger from freezing of exposed flesh within one minute.

GREAT DANGER
Flesh may freeze within 30 seconds.

Trenchfoot and immersion foot may occur at any point on this chart.

* Developed by U.S. Army Research Institute of Environmental Medicine, Natick, MA.

Equivalent chill temperature requiring dry clothing to maintain core body temperature above 36°C (96.8°F) per cold stress TLV®.

Evaluation and Control

For exposed skin, continuous exposure should not be permitted when the air speed and temperature results in an equivalent chill of −32°C (−25.6°F). Superficial or deep local tissue freezing will occur only at temperatures below −1°C (30.2°F) regardless of wind speed.

At air temperatures of 2°C (35.6°F) or less, it is imperative that workers who become immersed in water or whose clothing becomes wet be immediately provided a change of clothing and be treated for hypothermia.

TLVs® recommended for properly clothed workers for periods of work at temperatures below freezing are shown in Table 3.

Special protection of the hands is required to maintain manual dexterity for the prevention of accidents:

1. If fine work is to be performed with bare hands for more than 10 to 20 minutes in an environment below 16°C (60.8°F), special provisions should be established for keeping the workers' hands warm. For this purpose, warm air jets, radiant heaters (fuel burner or electric radiator), or contact warm plates may be utilized. Metal handles of tools and control bars should be covered by thermal insulating material at temperatures below −1°C (30.2°F).

2. If the air temperature falls below 16°C (60.8°F) for sedentary, 4°C (39.2°F) for light, −7°C (19.4°F) for moderate work, and fine manual dexterity is not required, then gloves should be used by the workers.

To prevent contact frostbite, the workers should wear anticontact gloves.

1. When cold surfaces below −7°C (19.4°F) are within reach, a warning should be given to each worker to prevent inadvertent contact by bare skin.

2. If the air temperature is −17.5°C (0°F) or less, the hands should be protected by mittens. Machine controls and tools for use in cold conditions should be designed so that they can be handled without removing the mittens.

Provisions for additional total body protection are required if work is performed in an environment at or below 4°C (39.2°F). The workers should wear cold protective clothing appropriate for the level of cold and physical activity:

1. If the air velocity at the job site is increased by wind, draft, or artificial ventilating equipment, the cooling effect of the wind should be reduced by shielding the work area or by wearing an easily removable windbreak garment.

2. If only light work is involved and if the clothing on the worker may become wet on the job site, the outer layer of the clothing in use may be of a type impermeable to water. With more severe work under such conditions, the outer layer should be water repellent, and the outerwear should be changed as it becomes wetted. The outer garments should include provisions for easy ventilation in order to prevent wetting of inner layers by sweat. If work is done at normal temperatures or in a hot environment

before entering the cold area, the employee should make sure that clothing is not wet as a consequence of sweating. If clothing is wet, the employee should change into dry clothes before entering the cold area. The workers should change socks and any removable felt insoles at regular daily intervals or use vapor barrier boots. The optimal frequency of change should be determined empirically and will vary individually and according to the type of shoe worn and how much the individual's feet sweat.

3. If exposed areas of the body cannot be protected sufficiently to prevent sensation of excessive cold or frostbite, protective items should be supplied in auxiliary heated versions.

4. If the available clothing does not give adequate protection to prevent hypothermia or frostbite, work should be modified or suspended until adequate clothing is made available or until weather conditions improve.

5. Workers handling evaporative liquid (gasoline, alcohol or cleaning fluids) at air temperatures below 4°C (39.2°F) should take special precautions to avoid soaking of clothing or gloves with the liquids because of the added danger of cold injury due to evaporative cooling. Special note should be taken of the particularly acute effects of splashes of "cryogenic fluids" or those liquids with a boiling point that is just above ambient temperature.

Work–Warming Regimen

If work is performed continuously in the cold at an equivalent chill temperature (ECT) or below –7°C (19.4°F), heated warming shelters (tents, cabins, rest rooms, etc.) should be made available nearby. The workers should be encouraged to use these shelters at regular intervals, the frequency depending on the severity of the environmental exposure. The onset of heavy shivering, minor frostbite (frostnip), the feeling of excessive fatigue, drowsiness, irritability, or euphoria are indications for immediate return to the shelter. When entering the heated shelter, the outer layer of clothing should be removed and the remainder of the clothing loosened to permit sweat evaporation or a change of dry work clothing provided. A change of dry work clothing should be provided as necessary to prevent workers from returning to work with wet clothing. Dehydration, or the loss of body fluids, occurs insidiously in the cold environment and may increase the susceptibility of the worker to cold injury due to a significant change in blood flow to the extremities. Warm sweet drinks and soups should be provided at the work site to provide caloric intake and fluid volume. The intake of coffee should be limited because of the diuretic and circulatory effects.For work practices at or below –12°C (10.4°F) ECT, the following should apply:

1. The worker should be under constant protective observation (buddy system or supervision).

2. The work rate should not be so high as to cause heavy sweating that will result in wet clothing; if heavy work must be done, rest periods should be taken in heated shelters and opportunity for changing into dry clothing should be provided.

TABLE 3. TLVs® Work/Warm-up Schedule for a 4-Hour Shift ☆

Air Temperature—Sunny Sky		No Noticeable Wind		5 mph Wind		10 mph Wind		15 mph Wind		20 mph Wind	
°C (approx.)	°F (approx.)	Max. Work Period	No. of Breaks	Max. Work Period	No. of Breaks	Max. Work Period	No. of Breaks	Max. Work Period	No. of Breaks	Max. Work Period	No. of Breaks
−26° to −28°	−15° to −19°	(Norm. Breaks)	1	(Norm. Breaks)	1	75 min	2	55 min	3	40 min	4
−29° to −31°	−20° to −24°	(Norm. Breaks)	1	75min	2	55 min	3	40 min	4	30 min	5
−32° to −34°	−25° to −29°	75 min	2	55 min	3	40 min	4	30 min	5	Non-emergency work should cease	
−35° to −37°	−30° to −34°	55 min	3	40 min	4	30 min	5	Non-emergency work should cease			
−38° to −39°	−35° to −39°	40 min	4	30 min	5	Non-emergency work should cease					
−40° to −42°	−40° to −44°	30 min	5	Non-emergency work should cease							
−43° & below	−45° & below	Non-emergency work should cease									

see next page for NOTES

TLV®-PA

TLV®–PA

NOTES for Table 3:

1. Schedule applies to any 4-hour work period with moderate to heavy work activity, with warm-up periods of ten (10) minutes in a warm location and with an extended break (e.g., lunch) at the end of the 4-hour work period in a warm location. For Light-to-Moderate Work (limited physical movement): apply the schedule one step lower. For example, at –35°C (–30°F) with no noticeable wind (Step 4), a worker at a job with little physical movement should have a maximum work period of 40 minutes with 4 breaks in a 4-hour period (Step 5).

2. The following is suggested as a guide for estimating wind velocity if accurate information is not available:
   5 mph: light flag moves; 10 mph: light flag fully extended; 15 mph: raises newspaper sheet; 20 mph: blowing and drifting snow.

3. If only the wind chill cooling rate is available, a rough rule of thumb for applying it rather than the temperature and wind velocity factors given above would be: 1) special warm-up breaks should be initiated at a wind chill cooling rate of about 1750 W/m²; 2) all non-emergency work should have ceased at or before a wind chill of 2250 W/m² In general, the warmup schedule provided above slightly under-compensates for the wind at the warmer temperatures, assuming acclimatization and clothing appropriate for winter work. On the other hand, the chart slightly over-compensates for the actual temperatures in the colder ranges because windy conditions rarely prevail at extremely low temperatures.

4. TLVs® apply only for workers in dry clothing.

☆ Adapted from Occupational Health & Safety Division, Saskatchewan Department of Labour.

3. New employees should not be required to work fulltime in the cold during the first days of employment until they become accustomed to the working conditions and required protective clothing.

4. The weight and bulkiness of clothing should be included in estimating the required work performance and weights to be lifted by the worker.

5. The work should be arranged in such a way that sitting still or standing still for long periods is minimized. Unprotected metal chair seats should not be used. The worker should be protected from drafts to the greatest extent possible.

6. The workers should be instructed in safety and health procedures. The training program should include as a minimum instruction in:

   a. Proper rewarming procedures and appropriate first aid treatment.

   b. Proper clothing practices.

   c. Proper eating and drinking habits.

   d. Recognition of impending frostbite.

   e. Recognition of signs and symptoms of impending hypothermia or excessive cooling of the body even when shivering does not occur.

   f. Safe work practices.

Special Workplace Recommendations

Special design requirements for refrigerator rooms include the following:

1. In refrigerator rooms, the air velocity should be minimized as much as possible and should not exceed 1 meter/sec (200 fpm) at the job site. This can be achieved by properly designed air distribution systems.

2. Special wind protective clothing should be provided based upon existing air velocities to which workers are exposed.

Special caution should be exercised when working with toxic substances and when workers are exposed to vibration. Cold exposure may require reduced exposure limits.

Eye protection for workers employed out-of-doors in a snow and/or ice-covered terrain should be supplied. Special safety goggles to protect against ultraviolet light and glare (which can produce temporary conjunctivitis and/or temporary loss of vision) and blowing ice crystals should be required when there is an expanse of snow coverage causing a potential eye exposure hazard.

Workplace monitoring is required as follows:

1. Suitable thermometry should be arranged at any workplace where the environmental temperature is below 16°C (60.8°F) so that overall compliance with the requirements of the TLV® can be maintained.

2. Whenever the air temperature at a workplace falls below –1°C (30.2°F), the dry bulb temperature should be measured and recorded at least every 4 hours.

TLV®-PA

3. In indoor workplaces, the wind speed should also be recorded at least every 4 hours whenever the rate of air movement exceeds 2 meters per second (5 mph).

4. In outdoor work situations, the wind speed should be measured and recorded together with the air temperature whenever the air temperature is below −1°C (30.2°F).

5. The equivalent chill temperature should be obtained from Table 2 in all cases where air movement measurements are required; it should be recorded with the other data whenever the equivalent chill temperature is below −7°C (19.4°F).

Employees should be excluded from work in cold at −1°C (30.2°F) or below if they are suffering from diseases or taking medication which interferes with normal body temperature regulation or reduces tolerance to work in cold environments. Workers who are routinely exposed to temperatures below −24°C (−11.2°F) with wind speeds less than five miles per hour, or air temperatures below −18°C (0°F) with wind speeds above five miles per hour, should be medically certified as suitable for such exposures.

Trauma sustained in freezing or subzero conditions requires special attention because an injured worker is predisposed to cold injury. Special provisions should be made to prevent hypothermia and freezing of damaged tissues in addition to providing for first aid treatment.

HEAT STRESS AND HEAT STRAIN

The goal of this TLV® is to maintain body core temperature within + 1°C of normal (37°C). This core body temperature range can be exceeded under certain circumstances with selected populations, environmental and physiologic monitoring, and other controls.

More than any other physical agent, the potential health hazards from work in hot environments depends strongly on physiological factors that lead to a range of susceptibilities depending on the level of acclimatization. Therefore, professional judgment is of particular importance in assessing the level of heat stress and physiological heat strain to adequately provide guidance for protecting nearly all healthy workers with due consideration of individual factors and the type of work. Assessment of both heat stress and heat strain can be used for evaluating the risk to worker safety and health. A decision making process is suggested in Figure 1. The exposure guidance provided in Figures 1 and 2 and in the associated *Documentation* of the TLV® represents conditions under which it is believed that nearly all heat acclimatized, adequately hydrated, unmedicated, healthy workers may be repeatedly exposed without adverse health effects. The Action Limit (AL) is similarly protective of unacclimatized workers and represents conditions for which a heat stress management program should be considered. While not part of the TLV®, elements of a heat stress management program are offered. The exposure guidance is not a fine line between safe and dangerous levels.

Heat Stress is the net heat load to which a worker may be exposed from the combined contributions of metabolic heat, environmental factors, (i.e., air temperature, humidity, air movement, and radiant heat), and clothing requirements. A mild or moderate heat stress may cause discomfort and may adversely affect performance and safety, but it is not harmful to health. As the heat stress approaches human tolerance limits, the risk of heat-related disorders increases.

Heat Strain is the overall physiological response resulting from heat stress. The physiological responses are dedicated to dissipating excess heat from the body.

Acclimatization is a gradual physiological adaptation that improves an individual's ability to tolerate heat stress. Acclimatization requires physical activity under heat-stress conditions similar to those anticipated for the work. With a recent history of heat-stress exposures of at least two continuous hours (e.g., 5 of the last 7 days to 10 of 14 days), a worker can be considered acclimatized for the purposes of the TLV®. Its loss begins when the activity under those heat stress conditions is discontinued, and a noticeable loss occurs after four days and may be completely lost in three to four weeks. Because acclimatization is to the level of the heat stress exposure, a person will not be fully acclimatized to a sudden higher level; such as during a heat wave.

The decision process illustrated in Figure 1, should be started if (1) a qualitative exposure assessment indicates the possibility of heat stress, (2) there are reports of discomfort due to heat stress, or (3) professional judgment indicates heat stress conditions.

Section 1: *Clothing.* Ideally, free movement of cool, dry air over the skin's surface maximizes heat removal by both evaporation and convection.

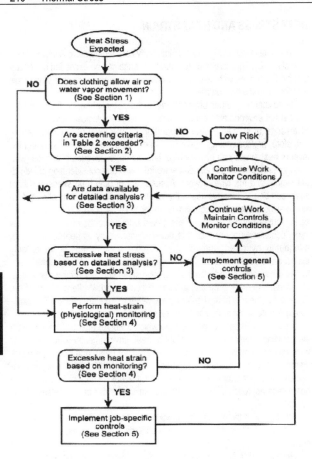

FIGURE 1. Evaluating heat stress and strain.

TLV®–PA

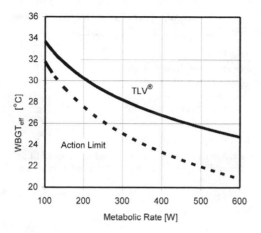

FIGURE 2. TLV® (solid line) and Action Limit (broken line) for heat stress. WBGT$_{eff}$ is the measured WBGT plus the Clothing-Adjustment Factor.

Evaporation of sweat from the skin is the predominant heat removal mechanism. Water-vapor-impermeable, air-impermeable, and thermally insulating clothing, as well as encapsulating suits and multiple layers of clothing, severely restrict heat removal. With heat removal hampered by clothing, metabolic heat may produce excessive heat strain even when ambient conditions are considered cool.

Figure 1 requires a decision about clothing and how it might affect heat loss. The WBGT-based heat exposure assessment was developed for a traditional work uniform of a long-sleeve shirt and pants. If the required clothing is adequately described by one of the ensembles in Table 1 or by other available data, then the "YES" branch is selected.

If workers are required to wear clothing not represented by an ensemble in Table 1, then the "NO" branch should be taken. This decision is especially applicable for clothing ensembles that are 1) totally encapsulating suits or 2) multiple layers where no data are available for adjustments. For these kinds of ensembles, Table 2 is not a useful screening method to determine a threshold for heat-stress management actions and some risk must be assumed. Unless a detailed analysis method appropriate to the clothing requirements is available, physiological and signs/symptoms monitoring described in Section 4 and Table 4 should be followed to assess the exposure.

Section 2: *Screening Threshold Based on Wet-Bulb Globe Temperature (WBGT).* The WBGT offers a useful first order index of the environmental contribution to heat stress. It is influenced by air temperature, radiant heat, air movement, and humidity. As an approximation, it does not fully account for all the interactions between a person and the environment and cannot account for special conditions such as heating from a radiofrequency/microwave source.

TABLE 1. Clothing-Adjustment Factors for Some Clothing Ensembles*

Clothing Type	Addition to WBGT [°C]
Work clothes (long sleeve shirt and pants)	0
Cloth (woven material) coveralls	0
Double-layer woven clothing	3
SMS polypropylene coveralls	0.5
Polyolefin coveralls	1
Limited-use vapor-barrier coveralls	11

*These values must not be used for completely encapsulating suits, often called Level A. Clothing Adjustment Factors cannot be added for multiple layers. The coveralls assume that only modesty clothing is worn underneath, not a second layer of clothing.

WBGT values are calculated using one of the following equations:

With direct exposure to sunlight:
$$WBGT_{out} = 0.7\, T_{nwb} + 0.2\, T_g + 0.1\, T_{db}$$

Without direct exposure to the sun:
$$WBGT_{in} = 0.7\, T_{nwb} + 0.3\, T_g$$

where:

T_{nwb} = natural wet-bulb temperature (sometimes called NWB)
T_g = globe temperature (sometimes called GT)
T_{db} = dry-bulb (air) temperature (sometimes called DB)

Because WBGT is only an index of the environment, the screening criteria are adjusted for the contributions of work demands and clothing. Table 2 provides WBGT criteria suitable for screening purposes. For clothing ensembles listed in Table 1, Table 2 can be used when the clothing adjustment factors are added to the environmental WBGT.

To determine the degree of heat stress exposure, the work pattern and demands must be considered. If the work (and rest) is distributed over more than one location, then a time-weighted average WBGT should be used for comparison to Table 2 limits.

As metabolic rate increases (i.e., work demands increase), the criteria values in the table decrease to ensure that most workers will not have a core body temperature above 38°C. Correct assessment of work rate is of equal importance to environmental assessment in evaluating heat stress. Table 3 provides broad guidance for selecting the work rate category to be used in Table 2. Often there are natural or prescribed rest breaks within an hour of work, and Table 2 provides the screening criteria for three allocations of work and rest.

Based on metabolic rate category for the work and the approximate proportion of work within an hour, a WBGT criterion can be found in Table 2 for

TABLE 2. Screening Criteria for TLV® and Action Limit for Heat Stress Exposure

Allocation of Work in a Cycle of Work and Recovery	TLV® (WBGT values in °C)				Action Limit (WBGT values in °C)			
	Light	Moderate	Heavy	Very Heavy	Light	Moderate	Heavy	Very Heavy
75 to 100%	31.0	28.0	–	–	28.0	25.0	–	–
50 to 75%	31.0	29.0	27.5	–	28.5	26.0	24.0	–
25 to 50%	32.0	30.0	29.0	28.0	29.5	27.0	25.5	24.5
0 to 25%	32.5	31.5	30.5	30.0	30.0	29.0	28.0	27.0

Notes:

- See Table 3 and the *Documentation* for work demand categories.
- WBGT values are expressed to the nearest 0.5 °C.
- The thresholds are computed as a TWA-Metabolic Rate where the metabolic rate for rest is taken as 115 W and work is the representative (mid-range) value of Table 3. The time base is taken as the proportion of work at the upper limit of the percent work range (e.g., 50% for the range of 25 to 50%).
- If work and rest environments are different, hourly time-weighted averages (TWA) WBGT should be calculated and used. TWAs for work rates should also be used when the work demands vary within the hour, but note that the metabolic rate for rest is already factored into the screening limit.
- Values in the table are applied by reference to the "Work-Rest Regimen" section of the *Documentation* and assume 8-hour workdays in a 5-day workweek with conventional breaks as discussed in the *Documentation*. When workdays are extended, consult the "Application of the TLV®" section of the *Documentation*.
- Because of the physiological strain associated with Heavy and Very Heavy work among less fit workers regardless of WBGT, criteria values are not provided for continuous work and for up to 25% rest in an hour for Very Heavy. The screening criteria are not recommended, and a detailed analysis and/or physiological monitoring should be used.
- Table 2 is intended as an initial screening tool to evaluate whether a heat stress situation may exist (according to Figure 1) and thus, the table is more protective than the TLV® or Action Limit (Figure 2). Because the values are more protective, they are not intended to prescribe work and recovery periods.

TABLE 3. Metabolic Rate Categories and the Representative Metabolic Rate with Example Activities

Category	Metabolic Rate [W] *	Examples
Rest	115	Sitting
Light	180	Sitting with light manual work with hands or hands and arms, and driving. Standing with some light arm work and occasional walking.
Moderate	300	Sustained moderate hand and arm work, moderate arm and leg work, moderate arm and trunk work, or light pushing and pulling. Normal walking.
Heavy	415	Intense arm and trunk work, carrying, shoveling, manual sawing; pushing and pulling heavy loads; and walking at a fast pace.
Very Heavy	520	Very intense activity at fast to maximum pace.

* The effect of body weight on the estimated metabolic rate can be accounted for by multiplying the estimated rate by the ratio of actual body weight divided by 70 kg (154 lb).

TLV®-PA

the TLV® and for the Action Limit. If the measured time-weighted average WBGT adjusted for clothing is less than the table value for the Action Limit, the NO branch in Figure 1 is taken, and there is little risk of excessive exposures to heat stress. If the conditions are above the Action Limit, but below the TLV®, then consider general controls described in Table 5. If there are reports of the symptoms of heat-related disorders such as fatigue, nausea, dizziness, and lightheadedness, then the analysis should be reconsidered.

If the work conditions are above the TLV® screening criteria in Table 2, then a further analysis is required following the YES branch.

Section 3: *Detailed Analysis.* Table 2 is intended to be used as a screening step. It is possible that a condition may be above the TLV® or Action Limit criteria provided in Table 2 and still not represent an exposure above the TLV® or the Action Limit. To make this determination, a detailed analysis is required. Methods are fully described in the *Documentation*, in industrial hygiene and safety books, and in other sources.

Provided that there is adequate information on the heat stress effects of the required clothing, the first level of detailed analysis is a task analysis that includes a time-weighted average of the Effective WBGT (environmental WBGT plus clothing adjustment factor) and the metabolic rate. Some clothing adjustment factors have been suggested in Table 1. Factors for other clothing ensembles appearing in the literature can be used in similar fashion following good professional judgment. The TLV® and Action Limit are shown in Figure 2.

The second level of detailed analysis would follow a rational model of heat stress, such as the International Standards Organization (ISO) Predicted Heat Strain (ISO 7933 2004; Malchaire et al., 2001). While a rational method (versus the empirically derived WBGT thresholds) is computationally more difficult, it permits a better understanding of the sources of the heat stress and is a means to appreciate the benefits of proposed modifications in the exposure. Guidance to the ISO method and other rational methods is described in the literature.

The screening criteria require the minimal set of data to make a determination. Detailed analyses require more data about the exposures. Following Figure 1, the next question asks about the availability of data for a detailed analysis. If these data are not available, the NO branch takes the evaluation to physiological monitoring to assess the degree of heat strain.

If the data for a detailed analysis are available, the next step in Figure 1 is the detailed analysis. If the exposure does not exceed the criteria for the Action Limit (or unacclimatized workers) for the appropriate detailed analysis (e.g., WBGT analysis, another empirical method, or a rational method), then the NO branch can be taken. If the Action Limit criteria are exceeded but the criteria for the TLV® (or acclimatized workers) in the detailed analysis are not exceeded, then implement general controls and continue to monitor the conditions. General controls include training for workers and supervisors, heat stress hygiene practices, and medical surveillance. If the exposure exceeds the limits for acclimatized workers in the detailed analysis, the YES branch leads to physiological monitoring as the only alternative to demonstrate that adequate protection is provided.

Section 4: *Heat Strain.* The risk and severity of excessive heat strain will vary widely among people, even under identical heat stress conditions. The normal physiological responses to heat stress provide an opportunity to monitor heat strain among workers and to use this information to assess the level of heat strain present in the workforce, to control exposures, and to assess the effectiveness of implemented controls. Table 4 provides guidance for acceptable limits of heat strain.

Following good industrial hygiene sampling practice, which considers likely extremes and the less tolerant workers, the absence of any of these limiting observations indicates acceptable management of the heat stress exposures. With acceptable levels of heat strain, the NO branch in Figure 1 is taken. Nevertheless, if the heat strain among workers is considered acceptable at the time, consideration of the general controls is recommended. In addition, periodic physiological monitoring should be continued to ensure acceptable levels of heat strain.

If limiting heat strain is found during the physiological assessments, then the YES branch is taken. This means that suitable job-specific controls should be implemented to a sufficient extent to control heat strain. The job-specific controls include engineering controls, administrative controls, and personal protection.

After implementation of the job-specific controls, it is necessary to assess their effectiveness and to adjust them as needed.

TLV®–PA

TABLE 4. Guidelines for Limiting Heat Strain

Monitoring heat strain and signs and symptoms of heat-related disorders is sound industrial hygiene practice, especially when clothing may significantly reduce heat loss. For surveillance purposes, a pattern of workers exceeding the heat strain limits is indicative of a need to control the exposures. On an individual basis, the limits represent a time to cease an exposure and allow for recovery.

One or more of the following measures may mark excessive heat strain, and an individual's exposure to heat stress should be discontinued when any of the following occur:

- Sustained (several minutes) heart rate is in excess of 180 bpm (beats per minute) minus the individual's age in years (e.g., 180 – age), for individuals with assessed normal cardiac performance; or
- Body core temperature is greater than 38.5°C (101.3°F) for medically selected and acclimatized personnel; or greater than 38°C (100.4°F) in unselected, unacclimatized workers; or
- Recovery heart rate at one minute after a peak work effort is greater than 120 bpm; or
- There are symptoms of sudden and severe fatigue, nausea, dizziness, or lightheadedness.

An individual may be at greater risk of heat-related disorders if:

- Profuse sweating is sustained over hours; or
- Weight loss over a shift is greater than 1.5% of body weight; or
- 24-hour urinary sodium excretion is less than 50 mmoles

EMERGENCY RESPONSE: If a worker appears to be disoriented or confused, suffers inexplicable irritability, malaise, or chills, the worker should be removed for rest in a cool location with rapidly circulating air and kept under skilled observation. Absent medical advice to the contrary, treat this as an emergency with immediate transport to a hospital. An emergency response plan is necessary.

— NEVER ignore anyone's signs or symptoms of heat-related disorders —

Section 5: *Heat Stress Management and Controls.* The elements of a heat stress management program including general and job-specific controls should be considered in the light of local conditions and the judgment of the industrial hygienist. The recommendation to initiate a heat stress management program is marked by 1) heat stress levels that exceed the Action Limit or 2) work in clothing ensembles that limit heat loss. In either case, general controls should be considered (Table 5).

Heat stress hygiene practices are particularly important because they reduce the risk that an individual may suffer a heat-related disorder. The key elements are fluid replacement, self-determination of exposures, health status monitoring, maintenance of a healthy lifestyle, and adjustment of expectations based on acclimatization state. The hygiene practices require the full cooperation of supervision and workers.

TABLE 5. Elements to Consider in Establishing a Heat Stress Management Program

Monitor heat stress (e.g., WBGT Screening Criteria in Table 2) and heat strain (Table 4) to confirm adequate control

General Controls

- Provide accurate verbal and written instructions, annual training programs, and other information about heat stress and strain
- Encourage drinking small volumes (approximately 1 cup) of cool, palatable water (or other acceptable fluid replacement drink) about every 20 minutes
- Encourage employees to report symptoms of heat-related disorders to a supervisor
- Encourage self-limitation of exposures when a supervisor is not present
- Encourage co-worker observation to detect signs and symptoms of heat strain in others
- Counsel and monitor those who take medications that may compromise normal cardiovascular, blood pressure, body temperature regulation, renal, or sweat gland functions; and those who abuse or are recovering from the abuse of alcohol or other intoxicants
- Encourage healthy lifestyles, ideal body weight and electrolyte balance
- Adjust expectations of those returning to work after absence from hot exposure situations and encourage consumption of salty foods (with approval of physician if on a salt-restricted diet)
- Consider pre-placement medical screening to identify those susceptible to systemic heat injury
- Monitor the heat stress conditions and reports of heat-related disorders

Job-Specific Controls

- Consider engineering controls that reduce the metabolic rate, provide general air movement, reduce process heat and water vapor release, and shield radiant heat sources, among others
- Consider administrative controls that set acceptable exposure times, allow sufficient recovery, and limit physiological strain
- Consider personal protection that is demonstrated effective for the specific work practices and conditions at the location

— **NEVER ignore anyone's signs or symptoms of heat-related disorders** —

TLV®–PA

In addition to general controls, appropriate job-specific controls are often required to provide adequate protection. During the consideration of job-specific controls, Table 2 and Figure 2, along with Tables 1 and 3, provide a framework to appreciate the interactions among acclimatization state, metabolic rate, work-rest cycles, and clothing. Among administrative controls, Table 4 provides acceptable physiological and signs/symptoms limits. The mix of job-specific controls can be selected and implemented only after a review of the demands and constraints of any particular situation. Once implemented, their effectiveness must be confirmed and the controls maintained.

The prime objective of heat stress management is the prevention of heat stroke, which is life-threatening and the most serious of the heat-related disorders. The heat stroke victim is often manic, disoriented, confused, delirious, or unconscious. The victim's body core temperature is greater than 40°C (104°F). If signs of heat stroke appear, aggressive cooling should be started immediately, and emergency care and hospitalization are essential. The prompt treatment of other heat-related disorders generally results in full recovery, but medical advice should be sought for treatment and return-to-work protocols. It is worth noting that the possibility of accidents and injury increases with the level of heat stress.

Prolonged increases in deep body temperatures and chronic exposures to high levels of heat stress are associated with other disorders such as temporary infertility (male and female), elevated heart rate, sleep disturbance, fatigue, and irritability. During the first trimester of pregnancy, a sustained core temperature greater than 39°C may endanger the fetus.

References

1. International Organization for Standardization (ISO): Ergonomics of the thermal environment – Analytical determination and interpretation of heat stress using calculation of the predicted heat strain. ISO 7933:2004. ISO, Geneva (2004).
2. Malchaire J; Piette A; Kampmann B; et al.: Development and validation of the predicted heat strain model. Ann Occup Hyg. 45(2):123–135 (2001).

2010 PHYSICAL AGENTS UNDER STUDY

The TLV® Physical Agents Committee solicits information, especially data, which may assist it in its deliberations regarding the following agents and issues. Comments and suggestions, accompanied by substantiating evidence in the form of peer-reviewed literature, should be forwarded in electronic format to The Science Group, ACGIH® (science@acgih.org). In addition, ACGIH® solicits recommendations for additional agents and issues of concern to the industrial hygiene and occupational health communities. Please refer to the ACGIH® TLV®/BEI® Development Process found on the ACGIH® website for a detailed discussion covering this procedure and methods for input to ACGIH® (http://www.acgih.org/TLV/DevProcess.htm).

The Under Study list is published each year by February 1 on the ACGIH® website (www.acgih.org/TLV/Studies.htm), in the ACGIH® Annual Reports, and later in the annual *TLVs® and BEIs®* book. In addition, the Under Study list is updated by July 31 into a two-tier list.

- Tier 1 entries indicate which chemical substances and physical agents **may** move forward as an NIC or NIE in the upcoming year, based on their status in the development process.
- Tier 2 consists of those chemical substances and physical agents that **will not** move forward, but will either remain on or be removed from the Under Study list for the next year.

This updated list will remain in two-tiers for the balance of the year. ACGIH® will continue this practice of updating the Under Study list by February 1 and establishing the two-tier list by July 31 each year.

The substances and issues listed below are as of January 1, 2010. *After this date, please refer to the ACGIH® website* (http://www.acgih.org/TLV/Studies.htm) *for the up-to-date list.*

1. Ergonomics
   - Hand-arm vibration
   - Localized fatigue
   - Whole-body vibration
2. Thermal Stress
   - Cold stress

TLV®-PA

2010
Biologically Derived
Airborne Contaminants

Contents

BDAC

2009 BIOAEROSOLS COMMITTEE

Joseph Torey Nalbone, MS, PhD, CIH — Chair
Paula H. Vance, SM(ASCP), SM(NRM) — Vice Chair
Michael S. Crandall, CIH, MSEE
Michael L. Muilenberg, MS
Francis (Bud) J. Offermann, CIH, PE
Carol Y. Rao, CIH, ScD, MS

BDAC

INTRODUCTION TO THE BIOLOGICALLY DERIVED AIRBORNE CONTAMINANTS

Biologically derived airborne contaminants include bioaerosols (airborne particles composed of or derived from living organisms) and volatile organic compounds that organisms release. Bioaerosols include microorganisms (i.e., culturable, nonculturable, and dead microorganisms) and fragments, toxins, and particulate waste products from all varieties of living things. Biologically derived contaminants are ubiquitous in nature and may be modified by human activity. Humans are repeatedly exposed, day after day, to a wide variety of such materials.

TLVs® exist for certain substances of biological origin, including cellulose; some wood, cotton, flour and grain dusts; nicotine; pyrethrum; starch; subtilisins (proteolytic enzymes); sucrose; vegetable oil mist; and volatile compounds produced by living organisms (e.g., ammonia, carbon dioxide, ethanol, and hydrogen sulfide). However, for the reasons identified below, there are no TLVs® against which to compare environmental air concentrations of most materials of biological origin.

ACGIH® has developed and separately published guidance on the assessment, control, remediation, and prevention of biologically derived contamination of indoor environments.[1] Indoor biological contamination is defined as the presence of a) biologically derived aerosols, gases, and vapors of a kind and concentration likely to cause disease or predispose humans to disease; b) inappropriate concentrations of outdoor bioaerosols, especially in buildings designed to prevent their entry; or c) indoor microbial growth and remnants of biological growth that may become aerosolized and to which humans may be exposed. The term "biological agent" refers to a substance of biological origin that is capable of producing an adverse effect, e.g., an infection or a hypersensitivity, irritant, inflammatory, or other response.

The ACGIH®-recommended approach to assessing and controlling bioaerosol exposures relies on visually inspecting buildings, assessing occupant symptoms, evaluating building performance, monitoring potential environmental sources, and applying professional judgment. The published guidance provides background information on the major groups of bioaerosols, including their sources and health effects, and describes methods to collect, analyze, and interpret bioaerosol samples from potential environmental sources. Occasionally, environmental monitoring detects a single or predominating biological contaminant. More commonly, monitoring reveals a mixture of many biologically derived materials, reflecting the diverse and interactive nature of indoor microenvironments. Therefore, environmental sampling for bioaerosols should be conducted only following careful formulation of testable hypotheses about potential bioaerosol sources and mechanisms by which workers may be exposed to bioaerosols from these sources. Even when investigators work from testable hypotheses and well-formulated sampling plans, results from environmental bioaerosol monitoring may be inconclusive and occasionally misleading.

There are no TLVs® for interpreting environmental measurements of a) total culturable or countable bioaerosols (e.g., total bacteria or fungi); b) specific culturable or countable bioaerosols (e.g., *Aspergillus fumigatus*); c) infectious agents (e.g., *Legionella pneumophila* or *Mycobacterium tuberculosis*); or d) assayable biological contaminants (e.g., endotoxin, mycotoxin, antigens, or microbial volatile organic compounds) for the following reasons.

BDAC

A. **Total culturable or countable bioaerosols.** Culturable bioaerosols are those bacteria and fungi that can be grown in laboratory culture. Such results are reported as the number of colony-forming units (CFU). Countable bioaerosols are those pollen grains, fungal spores, bacterial cells, and other material that can be identified and counted by microscope. A general TLV® for culturable or countable bioaerosol concentrations is not scientifically supportable because of the following:

1. Culturable microorganisms and countable biological particles do not comprise a single entity, i.e., bioaerosols in occupational settings are generally complex mixtures of many different microbial, animal, and plant particles.

2. Human responses to bioaerosols range from innocuous effects to serious, even fatal, diseases, depending on the specific material involved and workers' susceptibility to it. Therefore, an appropriate exposure limit for one bioaerosol may be entirely inappropriate for another.

3. It is not possible to collect and evaluate all bioaerosol components using a single sampling method. Many reliable methods are available to collect and analyze bioaerosol materials. However, different methods of sample collection and analysis may result in different estimates of culturable and countable bioaerosol concentrations.

4. At present, information relating culturable or countable bioaerosol concentrations to health effects is generally insufficient to describe exposure–response relationships.

B. **Specific culturable or countable bioaerosols other than infectious agents.** Specific TLVs® for individual culturable or countable bioaerosols have not been established to prevent hypersensitivity, irritant, or toxic responses. At present, information relating culturable or countable bioaerosol concentrations to health effects consists largely of case reports and qualitative exposure assessments. The data available are generally insufficient to describe exposure–response relationships. Reasons for the absence of good epidemiologic data on such relationships include the following:

1. Most data on concentrations of specific bioaerosols are derived from indicator measurements rather than from measurements of actual effector agents. For example, investigators use the air concentration of culturable fungi to represent exposure to airborne fungal antigens. In addition, most measurements are from either area or source samples. These monitoring approaches are less likely to reflect human exposure accurately than would personal sampling for actual effector agents.

2. Bioaerosol components and concentrations vary widely within and among different occupational and environmental settings. Unfortunately, replicate sampling is uncommon in bioaerosol assessments. Further, the most commonly used air-sampling devices for indoor monitoring are designed to collect "grab" samples over relatively short time intervals. Measurements from single, short-term grab samples may be orders of magnitude higher or lower than long-term average concentrations and are unlikely to represent workplace exposures accurately. Some organisms and sources release aerosols as "concentration bursts," which may only rarely be detected by limited grab sampling. Nevertheless, such episodic bioaerosol releases may produce significant health effects.

3. In studies of single workplaces, the number of persons affected by exposure to biological agents may be small if contamination is localized, thereby affecting only a fraction of the building occupants. However, data from different studies can seldom be combined to reach meaningful numbers of test subjects because the specific types of biological agents responsible for bioaerosol-related illnesses are diverse and often differ from study to study. These factors contribute to the low statistical power common in evaluations of cause–effect relationships between exposures to specific biological agents and building-related health complaints.

C. **Infectious agents.** Human dose–response data are available for only a few infectious bioaerosols. At present, air-sampling protocols for infectious agents are limited and suitable primarily for research endeavors. In most routine exposure settings, public health measures, such as immunization, active case finding, and medical treatment, remain the primary defenses against infectious bioaerosols. Facilities associated with increased risks for transmission of airborne infectious diseases (e.g., microbiology laboratories, animal-handling facilities, and health-care settings) should employ engineering controls to minimize air concentrations of infectious agents. Further, such facilities should consider the need for administrative controls and personal protective equipment to prevent the exposure of workers to these bioaerosols.

D. **Assayable biological contaminants.** Assayable, biologically derived contaminants (e.g., endotoxin, mycotoxins, antigens, and volatile organic compounds) are microbial, animal, or plant substances that can be detected using chemical, immunological, or biological assays. Evidence does not yet support TLVs® for any of these substances. However, assay methods for certain common airborne antigens and endotoxin are steadily improving, and field validation of these assays is also progressing. Dose–response relationships for some assayable bioaerosols have been observed in experimental studies and occasionally in epidemiologic surveys. Therefore, exposure limits for certain assayable, biologically derived, airborne contaminants may be appropriate in the future. In addition, innovative molecular techniques are becoming available for specific bioaerosols currently detectable only by culture or counting.

ACGIH® actively solicits information, comments, and data in the form of peer-reviewed literature on health effects associated with bioaerosol exposures in occupational and related environments that may help the Bioaerosols Committee evaluate the potential for proposing exposure guidelines for selected biologically derived airborne contaminants. Such information should be sent, preferably in electronic format, to The Science Group, ACGIH® (science@acgih.org).

Reference

1. ACGIH®: Bioaerosols: Assessment and Control. JM Macher, Ed; HM Ammann, HA Burge, DK Milton, and PR Morey, Asst. Eds. ACGIH®, Cincinnati, OH (1999).

BIOLOGICALLY DERIVED AGENTS UNDER STUDY

The Bioaerosols Committee solicits information, especially data, which may assist it in the establishment of TLVs® for biologically derived airborne contaminants. Comments and suggestions, accompanied by substantiating evidence in the form of peer-reviewed literature, should be forwarded in electronic format to The Science Group, ACGIH® (science@acgih.org).

The substances and issues listed below are as of January 1, 2010. *After this date, please refer to the ACGIH® website (http://www.acgih.org/TLV/Studies.htm) for the up-to-date list.*

Agents

gram negative bacterial endotoxin
(1-3) beta, D-glucan

CAS NUMBER INDEX

CAS NUMBER INDEX

CAS

CAS NUMBER INDEX

CAS NUMBER INDEX

CAS NUMBER INDEX

CAS NUMBER INDEX

CAS

CAS NUMBER INDEX

CAS NUMBER INDEX

CAS NUMBER INDEX

CAS NUMBER INDEX

CAS

CAS NUMBER INDEX

CAS

CAS NUMBER INDEX

CAS NUMBER INDEX

CAS NUMBER INDEX

CAS NUMBER INDEX

CAS

CAS NUMBER INDEX

9006-04-6	Natural rubber latex
9014-01-1	Bacillus subtilis [see Subtilisins]
10024-97-2	Nitrous oxide
10025-67-9	Sulfur monochloride
10025-87-3	Phosphorus oxychloride
10026-13-8	Phosphorus pentachloride
10028-15-6	Ozone
10035-10-6	Hydrogen bromide
10043-35-3	Boric acid
10049-04-4	Chlorine dioxide
10102-43-9	Nitric oxide
10102-44-0	Nitrogen dioxide
10210-68-1	Cobalt carbonyl
10294-33-4	Boron tribromide
11097-69-1	Chlorodiphenyl (54% chlorine)
11103-86-9	Zinc potassium chromate
12001-26-2	Mica
12001-28-4	Crocidolite [see Asbestos]
12001-29-5	Chrysotile [see Asbestos]
12079-65-1	Manganese cyclopentadienyl tricarbonyl
12108-13-3	2-Methylcyclopentadienyl manganese tricarbonyl
12125-02-9	Ammonium chloride fume
12172-73-5	Amosite [see Asbestos]
12179-04-3	Borates, tetra, sodium salts, pentahydrate
12185-10-3	Phosphorus (yellow)
12604-58-9	Ferrovanadium
13071-79-9	Terbufos
13121-70-5	Cyhexatin [Tricyclohexyltin hydroxide]
13149-00-3	Hexahydrophthalic anhydride, cis-isomer
13463-39-3	Nickel carbonyl
13463-40-6	Iron pentacarbonyl
13463-67-7	Titanium dioxide
13466-78-9	Δ-3-Carene [see Turpentine]
13494-80-9	Tellurium
13530-65-9	Zinc chromate
13765-19-0	Calcium chromate
13838-16-9	Enflurane
14166-21-3	Hexahydrophthalic anhydride, trans-isomer
14464-46-1	Silica, crystalline — cristobalite
14484-64-1	Ferbam
14807-96-6	Talc (nonasbestos form)
14808-60-7	Silica, crystalline — quartz
14857-34-2	Dimethylethoxysilane

CAS

CAS NUMBER INDEX

CAS

CAS NUMBER INDEX

NOTES

NOTES

<u>NOTES</u>

NOTES

NOTES

For reference

Not to be taken from the room.